Robert Schoenfeld

The Chemist's English

Distribution:

VCH Verlagsgesellschaft, Postfach 1260/1280, D-6940 Weinheim
(Federal Republic of Germany)

USA and Canada: VCH Publishers, 303 N.W. 12th Avenue, Deerfield Beach,
FL 33442-1705 (USA)

ISBN 3-527-26309-8 (VCH Verlagsgesellschaft)
ISBN 0-89573-436-2 (VCH Publishers)

Robert Schoenfeld

The Chemist's English

Robert Schoenfeld
Managing Editor
Australian Journal of Chemistry
314 Albert Street
East Melbourne
Australia 3002

Editoral Director: Dr. Hans F. Ebel
Production Manager: Dipl.-Ing. (FH) Hans Jörg Maier

Library of Congress Card No. 85-17858

CIP-Kurztitelaufnahme der Deutschen Bibliothek

Schoenfeld, Robert:
The Chemist's English / Robert Schoenfeld. – Weinheim; Deerfield Beach, Fl.: VCH, 1985.
  ISBN 3-527-26309-8 (Weinheim)
  ISBN 0-89573-436-2 (Deerfield Beach)

Composition: Filmsatz Unger, D-6940 Weinheim
Printing: betz-druck gmbh, D-6100 Darmstadt 12
Bookbinding: Josef Spinner, Großbuchbinderei GmbH, D-7583 Ottersweier
Printed in the Federal Republic of Germany

For Charles Hurd

Preface

I imagine you, dear reader, in a bookshop, having just picked this little volume off the shelf. You turn to this Preface to find out whether here is something worth buying. Well, I am an honest man; I won't sell my goods under false pretenses. This book does not set out to *teach* you to write good English. The skill of good writing cannot be taught, just as the skill of doing research cannot be taught.

All right, the book will teach you to avoid a number of annoying errors that may cause delay in getting your work published, and for this merit alone you may consider the book is worth having. But any such merit is only a by-product. I do not set out to give you a recipe for good scientific writing; I want to give you an *appetite* for good English.

I said that the skill of doing research cannot be taught. Your professors can show you how to acquire a data base and how to manipulate it, and how to gain mastery of the gleaming tools of scientific research. But you need more, and nobody can give you that: you need the desire to explain what has hitherto remained unexplained. Unless you have that you will remain an honest data-gatherer all your life; but research is something else.

Much the same is true of writing. You need the data base: the vocabulary. Then you need certain manipulative skills, such as shaping a sentence by putting a noun phrase and a verb phrase into it and getting the two to agree. All this can be achieved by teaching at a very low level; in fact, if the language in question happens to be your native one, you may reach 90% of your speaking potential and 80% of your writing potential in your teens. So, certainly, there is a teachable component in language. But it is not enough; to become a good writer you must also have an intense *desire* to communicate. And there is no recipe or algorithm that will give you this desire.

Perhaps this book can help. It is written, above all, for scientists, and it makes two assumptions: (1) you have something to say; (2) you are inhibited by fear of making mistakes.

I am confident I can remove that fear. First, by making you laugh; you will find that every misfortune that has happened to you has already happened to others, sometimes with pretty amusing results, and yet these others have survived the experience. Second, by analysing these mistakes. You will protest that others have already tried to do that for you, with the result that you have become more inhibited than ever. But those others have used the traditional methods of the grammarian, whereas I am a chemist talking to chemists. We shall use the chemist's tools: we shall treat sentences as though they were molecules, and the words as though they were atoms. We shall then perform such familiar operations as determining how these atoms are bonded together, how strong the bonds are, and what contribution each part of the molecule makes to the general stability.

Let me sum up what this book sets out to do. I do not promise to teach you good writing. I promise to keep you entertained and hope to give you confidence, so that you will want to sit down as soon as possible at your writing desk, eager to tell your public about your latest results. Who knows, I might even interest you in the laws of language. Did you know that the science of language is still in its infancy? For centuries it was believed that the traditional study of grammar, in the footsteps of Roman scholars such as (I have just looked up the name in an encyclopedia) Priscian, would yield all the facts about language that could be known. But then came the computer age and people tried to translate from one language to another by electronic means. It promptly became apparent that very little was known about the laws of language. For instance, in an English sentence, by what process do you find the verb? (Your dictionary will tell you that *sentence, instance, process* and *find* could be either verbs or nouns.)

The other day I found out (but inevitably many others must have commented on the fact before me) that about half of the human race puts the adjective before the noun, and half behind. Now why should that be? Surely one order is more efficient than the other, and should have prevailed. That's linguistics for you, one fascinating but yet unexplained fact after another.

I hope you will want to read the collection of essays that follows. But perhaps you draw back; you wonder whether you might not be better off with a traditional book on English style, written by a professor of English. There are a number of such books on the market, by distinguished authors, and I have often turned their pages in admiration, but in all my years as an editor I have never found a chemist whose prose style has been improved by reading them. In our library there is a particularly famous one, "English

Prose Style" by Herbert Read (G. Bell & Sons, London); it has gone through numerous reprintings and these have invariably been greeted with enthusiasm by critics in literary magazines. In the introduction, the author develops the reasonable thesis that good writing can be learned by reading consistently good writers. Let me cite one of his sentences:

> Take this test only: of how many writers, in the search for an appropriate and representative passage, could we trust to the offering of any page we opened at?

I don't want to be sardonic. I shall just say that this sentence reminds me of a highly strained molecule. Is that the kind of prose you want to write your next paper in?

It was precisely because I felt that the standard texts were not helpful to chemists that this collection of essays came into being. I published Parts 1 – 30 in the *Proceedings of the Royal Australian Chemical Institute* and its successor *Chemistry in Australia,* and Parts 31 – 33 in *CHEMTECH.* I thank the Royal Australian Chemical Institute and the American Chemical Society for allowing me to make use of this material. I am the editor of the *Australian Journal of Chemistry* and in some of the articles allude to specific Australian problems; I hope this will interest rather than annoy my readers.

The community of research chemists in Australia and New Zealand is not numerous; we meet at congresses; everyone seems to know all the others and to be on first name terms with them. This explains the tone of my articles: I imagined myself chatting to a group of friends. When I began to get correspondence from overseas I found that readers from outside our circle of acquaintance also liked this type of presentation; thus I have not changed it in this book version. I have taken care in revising, though, to make sure that these articles are accessible not only to chemists but to scientists of other disciplines as well.

Have fun with my articles. If you agree with my opinions, this will flatter my vanity; if you disagree, prove me to be in error by fashioning this or that particularly taut and effective paragraph in your next article or book or report. The best science has been done, and the best prose has been written, by those confident enough to disagree with authority.

Melbourne, Spring 1985   R. Schoenfeld

Contents

XII   *Contents*

1   To Get Acquainted . . .

I am an editor of the *Australian Journal of Chemistry*. Some years ago, I began to write a series of articles on some peculiarities of scientific English that I found amusing, hoping to share my amusement with others. In this I was successful; I received invitations to lecture at universities and research institutes; and during the convivial occasions that followed, my hosts confessed why they found it difficult to write. "All these rules of grammar", they said,"frighten us off".

So let this be the first message of this book: Don't be frightened of grammar. When you sit down to write your paper or thesis or report, your most dangerous enemy is not the split infinitive – it is ambiguity. A split infinitive is very often acceptable anyway, but where it needs correcting it can be corrected by a copy-editor. However, the copy-editor, unless he is a mindreader, cannot correct an ambiguity. So, even if you are not a smooth writer, don't sit there staring at the blank page; get your facts down first and fix up the dangling participles afterwards.

And why are you so afraid of grammar, anyway? You are a graduate in chemistry or a related science; that means your mind is disciplined enough to understand and organize large sets of difficult concepts. Why should grammar (I speak of traditional or schoolmaster's grammar here) be beyond your reach? I shall prove to you, here and now, that grammar is a science inferior to chemistry.

Let us perform a thought experiment: we assume that at a given university two research workers, a traditional grammarian and a chemist, begin their work at the same hour. Their work is at first of the same order: they each study a population of experimental data. Pretty soon both these learned men begin to observe certain patterns and regularities in their populations, and that permits them, by deductive reasoning, to formulate a scientific law. For the grammarian, this may take the form of "All nouns of multitude preceded by the indefinite article take the plural". The chemist will say "All ketones react with phenylhydrazine to give a phenylhydrazone".

So far so good; neither man has outperformed the other. But now they bend their heads over their work again and it is not long before they find exceptions to their laws. The grammarian finds that there are nouns of multitude, such as *set,* that capriciously prefer the singular. The chemist discovers that there are ketones that will not form phenylhydrazones no matter how hard one tries. At this point the difference between the sciences begins to emerge. The grammarian cannot take his research any deeper. All he can do is to amend his law to: "All nouns of multitude ... take the plural *except for* ..." For him, an exception to his law is a setback. To the chemist, though, it is a signal that he is about to find something new. He wants to know why his wayward ketones misbehave, and comes up with new insights into the geometry of molecules or the changes in electron distribution brought about by substituents. Instead of amending, and thereby weakening, his first law, he will replace it by a more subtle one beginning with "Ketones form phenylhydrazones only if ..." And having formulated this law, and having sent the story of its genesis to his favorite journal, our chemist will race back to his laboratory in an eager search for what the grammarian dreads most: exceptions.

I hope you are convinced: we chemists are smart enough to deal with problems of grammar. In fact, this will be one of the themes of this book: if we use the chemist's tools of deduction, we shall often discover that the laws of grammar are not disjointed statements lying side by side, but can be connected by a logical thread.

Have I cured you of your fear of your old schoolmaster's ghost? There is still another fear that may afflict you as you sit, pondering, in front of the blank page. You may have the feeling that knowing English is not enough, that the Chemist's English is somehow different from the language you learned at school, and that you are not at ease with this strange language.

To a certain extent this is true, but it is nothing to be afraid of. This book is meant to be accessible not only to native speakers of English, but also to non-natives sufficiently advanced to consider publishing in that language. But let us assume for the moment that you have been brought up on English. You have, of course, a feeling of complete mastery over Standard Spoken English (SSE) and you also feel at ease with Standard Written English (SWE), but Chemist's English (CE) is something else again, and at times you may find yourself translating a phrase from SWE to CE rather than writing directly in that language.

Is CE really a separate language, then? Legitimately, wo could call it a subspecies of SWE. There are significant differences between CE and SWE in sentence construction (CE has a far higher proportion of sentences with the verb in the passive form) and vocabulary (in CE words with emotional undertones are taboo, and pronouns of the first person are avoided). There are also slight differences in grammar.

Let us take a sentence in SSE and turn it into SWE and CE. For example, you may be chatting with a friend in the laboratory, and you say in SSE:

> We had an idea that the hydroxy group was tertiary, so we stewed the compound up with acetic anhydride.   (1)

Next we find you writing to a colleague, in SWE. The sentence now becomes:

> We suspected the hydroxy group was tertiary, so we heated the compound with acetic anhydride.   (2)

Now to translate into CE. We have been trained to avoid "we", and the word "suspect" is suspect, so we write:

> In order to determine whether the hydroxy group was tertiary, the compound was heated with acetic anhydride.   (3)

Let us note in passing that there is a decrease in signal-to-noise ratio through the series [the SSE sentence (1) has implications of an impulsively formed hypothesis and of forcing conditions; these are progressively lost]. But the point I am trying to make is this: sentence (3) is acceptable CE, but would not be considered good SWE by most authorities.

The SWE grammarians' argument against (3) is as follows: The phrase "In order ... was tertiary" expresses a purpose, and whenever such a phrase precedes the principal sentence, that purpose is ascribed to the subject of that sentence. I have used non-technical language, and I hope you will get the point: if you look at sentence (3) again, you will see it literally means that the compound (not the authors, who are not mentioned) got curious about the nature of its hydroxy group. Let us abandon CE for an instant, and create the equivalent of (3) in the Sociologist's English:

> In order to determine their suitability for the chemistry course, the students underwent an aptitude test.   (4)

No doubt you will read (4) with a slight feeling of unease; it is a bad sentence because it does not say exactly what was meant. It seems as if the students themselves had clamored for the aptitude test, whereas in fact it was inflicted on them by the instructors.

[I should mention here that the trouble with sentences (3) and (4) does not arise when the phrase of purpose *follows* the verb of the principal sentence, presumably because in that case there is no room for confusion. "The solution was boiled to drive off HBr" is good CE and correct SWE, and a sentence such as "The wheat was sown early that year, to take advantage of the unusual weather" could occur at the beginning of a fastidious author's novel.]

We are left with the fact that sentence (3) is acceptable in CE whereas a meticulous editor might object to it in SWE. Perhaps I had better present my evidence concerning the acceptability of (3). On the day I decided to write this chapter, I had on my desk No. 9/10 of *J. Chem. Soc., Perkin Trans. 1,* 1972. A ten-minute search uncovered four relatives of sentence (3). I shall quote from p. 1165: "In an attempt to generate the cation . . . by an alternative route, the diazonium sulphate . . . was heated in methanol." Similar sentences occur on pp. 1205, 1223 and 1232.

I took this to prove my point, as the *J. Chem. Soc.* is a carefully edited journal. I should still mention that I was less successful with No. 8 of Vol. 94 of *J. Am. Chem. Soc.,* which I picked up next. I found only one sentence of type (3) (top of p. 2844), but there were several instances of authors following up an "in order to . . ." phrase quite correctly with ". . . we heated . . .", ". . . we determined . . ." etc. Such sentences, of course, are perfect SWE because the purpose is correctly seen to be that of the subject. It would obviously be irresponsible to theorize from such a small sample, but the thought did cross my mind that contributors to American journals might be less inhibited about the use of the first person plural, and thus less likely to get themselves into grammatical mischief.

Be that as it may, we remain with the conclusion that CE and SWE are drifting slightly apart, and the question arises whether we should worry about this. Some very distinguished people do worry, and occasionally I read articles in which chemists are asked, in substance, to write their papers in SWE rather than CE. Personally, I agree with these eminent writers, but this book is concerned with the Chemist's English now prevailing, not with some ideal language.

Under prevailing customs, then, the CE sentence we have analysed would be considered acceptable in any scientific journal I can think of.

Very early in my career I might have been zealot enough to move the phrase of purpose to the back of the sentence, but every editor soon learns to be as tolerant as possible. However, prevailing customs also permit the authors — and this is the best solution, provided it is not done to excess — to use the first person plural whenever a sentence expresses something subjective like an opinion, a purpose, a decision.

Prevailing custom, alas, is still against the use of the first person singular. I admit there are moments when the use of this English one-letter word is contraindicated. Figure this one out:

> Analysis revealed that Cl and Br were present in appreciable amounts; but as stated in Part I I found I only in traces.   (5)

Thus a certain amount of care with the letter is certainly desirable, but if you are the sole author of your paper I see no harm in your using *I* in those places where, as a member of your group, you would have written *we*.

I realize I contradict myself here: just a moment ago I said I would simply record prevailing customs, and here I find myself advocating a change. But customs are progressively changing in the direction I desire, the change suggested is only marginal, and I hope I shall be forgiven if I try to push things along just a little. Certainly, in my journal, all you have to do is to promise not to write about atomic iodine, and I shall promise in turn not to put any road-block in the way of your legitimate little ego-trip!

2 The Search for the Missing Ablative

So you think you have problems with grammar? Be glad your were not born a Goth or a Phoenician. Some ancient laguages had grammatical forms of the most bizarre complexity. In the Basque language, if you alter the ending of a verb you thereby indicate that you stand in awe of the person you are addressing. In Hebrew, if you change a verb's vowels you thereby give it special emphasis.

No one is sorry that such odd refinements are missing from English grammar. We would much rather preface a verb with "respectfully" or "emphatically" than learn a whole new set of verb forms. But some of these ancient forms did have their uses. If you listen (as every scientist or technologist ought to, these days) to a moral philosopher, you might hear him complain about the fact that the English language does not have a gerundive. A gerundive, in Latin, was a suffix which, when tacked on to a verb, indicated that the action the verb described *ought* to be done *(nunc est bibendum, Carthago est delenda)*. In English, "to praise" has a *de facto* gerundive in "praiseworthy", but very few other verbs do, and our philosopher has to circumlocute to express thoughts that Seneca could have formulated with a few syllables.

Anyhow, we chemists do not miss the gerundive much (except those of us who have just established a pursueworthy hypothesis). But we certainly could do with a nice ablative. In fact, the absence of an ablative is one of the most annoying problems in the Chemist's English.

The ablative, in case you have forgotten, is a Latin case which works as follows. You make some clever change to the ending of a noun, and it thereupon becomes apparent that what-the-noun-means is the cause or the motive or the instrument of what the verb says is going on. Quintus killed Sextus (somebody always gets killed in these Latin exercises) *in* a rage, *with* a knife, *by* a stratagem, *by means of* slow poison — all the italicized prepositions Cicero was able to leave out, because he had studied his ablatives.

In the Slav languages, something like the ablative still survives, but all other modern languages have lost it. Why did this very useful grammatical form die out? No doubt its signal-to-noise ratio was too high for that imperfect decoding system, the human ear; there were some problems of ambiguity; and it was a bother having to remember a new suffix for each declension and for hordes of irregular nouns. So most modern languages, instead of altering the ending of a noun, place a preposition in front of it. In German, this preposition is *durch* (replaced occasionally by synonyms such as *vermittelst*), in French *par* (or *au moyen de* . . .), in Italian *per* (or *tramite* or *mediante* . . . .).

In English, we seem at first glance to have a particularly impressive array of synonymous prepositions: *with, by, by means of*. But closer inspection uncovers what is in fact a genuine weakness in scientific English. The prepositions are not interchangeable, that is, their meanings do not always overlap. Hence we must choose them with care and cannot always be certain of our choice. And custom is very capricious: we heat our solutions *with* a Bunsen burner, *by* ultraviolet radiation, *by means of* a heating mantle.

For Chemist's English there exists a rough rule of thumb: *with* is used for simple instruments, *by* for more complex ones and above all for abstract names of methods and techniques, *by means of* for yet more complex instruments and concepts. (Crystals are separated with a spatula, liquids by distillation, and vapours by means of the Hirakawa Smith-Jones 345 A Kromo-Graf apparatus.) "By means of" very often overlaps "by" and sometimes "with", in fact it comes nearest to being an all-purpose pseudo-ablative; but it is a cumbersome phrase and its constant repetition makes for leaden prose.

What proves that the problem is very real is the fact that chemists are still earnestly searching for a simple English word that will perform all the services rendered in other languages by *durch, par* and *per*. In my time as an editor I have seen three such words rise to prominence, with varying degrees of acceptance.

The first of these, and in my view the best qualified, was *through*. This comes, of course, from the same root as *durch* and I suspect it is no coincidence that it had its crest of popularity at a time when many standard textbooks were all-too-faithful translations from the German. *Through* in the sense we are discussing appears only very occasionally in the chemical literature now; it just has not passed the popularity test (perhaps the literal meaning is too readily present in the reader's mind) and we all feel that there is something odd about "The compounds were separated through chromatography". And yet this public rejection is somewhat unjust. *Through* as a synonym of *by means of* has a perfectly honorable history; its predecessor in Old English was used in precisely the manner of *durch*, and English Kings reigned *thurh godes fultome* before they became monarchs by the grace of God. This use has never died out; phrases such as "through sheer persistence" do not at all seem contrived. Hence, if chemical writers were to attempt a cautious and moderate revival of *through,* perhaps as an occasional substitute for *by means of,* they would thereby protect the Chemist's English from less desirable alternatives.

The second of these ablative-substitutes occupies a dominant position at the moment. This is *using* as employed in

The substance was analysed *using* mass spectrometry.   (1)

I am very much opposed to the use of this word, although I know full well that if an expression, no matter how badly conceived, gains wide acceptance, an editor's protests are to no avail. But let me, while I still have the English dictionaries on my side, explain my objections to the misuse (we shall see in a moment there is also a legitimate use) of *using.* The argument runs as follows: (i) *using* is listed in the dictionaries as a participle; (ii) a participle, if correctly used, is always attached to a noun or pronoun; (iii) *using* in sentence (1) is not so attached; (iv) therefore it is incorrectly used.

This situation has of course come about by the chemist's preference for sentences with passive verbs. Had the sentence run

We analysed the compound using mass spectrometry.   (2)

then the word *using,* being correctly attached to *we,* would have passed muster. [A careful author, in sentence (2), would insert a comma before *using,* to make it quite plain that *using* is attached to *we* and not to *compound.*]

When I first wrote about the missing ablative, in 1974, I was fairly forthright in my condemnation of sentence (1). Now, revising in 1984, I have to be much more prudent. All I can report at the moment is that sentence (1) is still incorrect in Standard Written English, but is tolerated by most editors of the Chemist's English.

I still live in hopes that this particular tolerance can be reversed. It still seems to me that sentence (1) cheapens the English language; in fact I feel so strongly about the matter that I shall return to it in the chapters on participles and prepositions. But I am aware that my chances in this campaign are small. It is far more likely that the misuse of *using* will spread from the Chemist's English to Standard Written English, and that then the English dictionaries, most of which record such developments uncritically, will list *using* as a preposition. With that my battle will be lost, and I must admit, in fairness, that the transmutation of participles into prepositions is a time-honored English process – think only of *pending, owing, according, except . . .*

Thus I must content myself with saying that your paper will impress discerning readers much more if you keep the cousins of sentence (1) out of it. Whenever this sentence tempts you, see whether you cannot replace *using* with *with*, or by *by*, and you may even consider *by means of*. If all these remedies fail, there is still sentence (2).

Most editors are only wary of *using,* but they genuinely dislike the third ablative-substitute, *via*, as badly used in "separated *via* chromatography". I must admit that, to my intense grief, this use of *via* has found its way into the otherwise very scholarly Webster's Dictionary. But members of reputable scientific publishing houses agree with Sir Ernest Gowers: "*Via* can only be properly used of the route; to apply it to the means of transport is a vulgarism." Translated into Chemist's English, this means: you can only use *via* for intermediates, not for techniques. Isolate a base via its picrate by all means, or proceed from an alcohol to an acid via an aldehyde. But don't try to determine the structure of these compounds via n.m.r. If you do, the reviewer may return your manuscript to you for correction, via the editor.

3   *Arguing with Authority*

Let us assume you have just written a paper or a report. The head of your department, or an internal referee, criticizes a word or a sentence, declaring it to be bad English. You disagree; perhaps you think he is too fussy, or perhaps you feel confident of having done what you were taught to do many years ago. How can you find out who is right?

In the first instance, your search for arguments will lead you to a dictionary. If you live in Australia as I do, or in Great Britain, your reference book is likely to be the Concise Oxford Dictionary, or COD. In many ways, this is a pity. Dictionaries can be classified very much like boxers, and it is true that in the heavyweight category the British entry, the Oxford English Dictionary, reigns as the unbeatable champion. But in the lightweight to light-heavyweight divisions, which matter most for our purposes, the position is quite different. British dictionaries seem mostly to be compiled for the use of schoolboys, or else for philologists, and ignore with lofty disdain the needs of typists, journalists, editors, and printers' readers. Thus the difficult questions of capitalization, and division of words at the end of a line, are carefully avoided, and the hyphenation is decidedly old-fashioned (to quote the COD). Still, the COD is a worthy competitor in the lightweight class (its virtues are legible type and tidy definitions). In the light-heavyweight class, though, the third edition of Webster's Third New International Dictionary is in my opinion a far better performer than its competitors (e. g. Shorter Oxford Dictionary, Random House).

Anyhow, any good dictionary from lightweight up will settle such simple disputes as the choice between *alternate* and *alternative,* or *fortunate* and *fortuitous.* It can also decide subtler questions. For instance, when preparing Chapter 2, I made certain that all leading dictionaries listed *using* only as the participle of the verb *to use,* and not in its own right as a preposition. Hence, I advised my readers to avoid, for the time being, such sentences as "The compounds were separated using chromatography".

If you possess instinctive dictionary-sense and a good knowledge of grammatical terms, your dictionary can help you to decide even more intricate questions. Suppose you have written: "The compound was reacted with diazomethane". Your critic objects, and maintains the sentence should read "The compound was made to react with diazomethane". The dictionary, after the entry 'react', shows the letters *v. i.* This settles the argument in favour of your critic — provided you know that *v. i.* stands for *verb intransitive,* and that you know what an intransitive verb is. (If you do not, you will find it made plain in Chapter 34.)

But a dictionary is designed to deal with only one word at a time, or at best small groups of words. If the dispute is about sentence structure, the cry will soon go up: "Let's look it up in Fowler".

A Dictionary (actually a collection of essays, alphabetically arranged) *of Modern English Usage* appeared in 1926, and through its instant and enduring success Henry Watson Fowler came to dominate the English language as only Dr Johnson had done before him. As time passed and the language changed, similar compendia appeared, but they all stood under his influence (the best known of these is Eric Partridge's *Usage and Abusage*). In 1965 (Fowler died in 1933) Clarendon Press brought out a Second Edition, sensibly revised and modernized by Sir Ernest Gowers. This is the book that should be on your writing desk today.*

But how does one ever find the answer to one's problem in *Fowler*? Some fortunate persons can find the appropriate essay instinctively, just as there are chemists who can locate a compound in Beilstein in seconds; others turn feverishly page after page and in the end retire defeated. The 1926 Fowler was intended to be read, I think, as a bedside book. The reader was meant to open the book at random; a system of cross-references would inevitably direct him to the essays dealing with fundamental matters. Gowers has made things easier for the reader by introducing a Classified Guide which leads the reader to the basic essays. But random reading is still the best way to get acquainted with Fowler; in your hour of need you will then be able to confound your critic by producing the argument-winning essay with nonchalant ease.

---

* Recently a number of handbooks on usage have appeared that no longer quite fit into the Fowler mould. The best I know is *The Careful Writer* by Theodore M. Bernstein (Atheneum: New York, 1966). The author is a first-rate journalist (former editor of the *New York Times*) and his writing is brisk, lucid, and completely free of headmasterly condescension.

And a very convincing essay it will be. Fowler's influence on the English language was almost in its entirety beneficial. He was the man who permitted us to ruthlessly split an infinitive whenever there was no simpler way out; he allowed us to choose the right preposition to end a sentence with; and he removed all restrictions from the word *whose* whose use when referring to inanimate objects the pedants had prohibited.

But even the most benevolent despot has his obsessions. Fowler had an obstinate belief that a distinction should be made between *due to* and *owing to. Due,* he maintained, was a participle and could only be used when linked to a noun; *owing* was a participle-turned-preposition and therefore free from this restriction. It was thus incorrect to say "Due to its low solubility, the magnetic properties of the compound could not be determined" (here *due* is unattached), but "The reactivity of the compound is due to its keto group" was correct (*due* here is attached to *reactivity*). *Owing to* could have been used in either sentence.

This is a battle that Fowler lost. Gowers, in the second edition, all but concedes defeat and even cites a passage where "the offending usage has indeed become literally part of the Queen's English". All chemists will commend Her Majesty for disregarding this absurd shibboleth; why indeed consider 'due' as a participle when the verb it is derived from has long disappeared from the language? I invite all authors to use *owing to* and *due to* interchangeably but might I at the same time make a plea for the return of the unfairly neglected *because of?*

The purists are about to lose another battle, but in this one my sentiments are on their side. They make a distinction between *expect* (foreseeing an event) and *anticipate* (foreseeing an event, and doing something about it before it happens). It is a pity this useful distinction should be lost, but most chemist-writers are no longer aware of it. Defenders of the distinction often make the point that expecting a marriage is not the same thing as anticipating it. But, at a time when social customs change, can we expect verbal customs to remain unchanged?

4 Defying the Dictionary

In the last chapter I paid a tribute to men like Fowler, Partridge and Gowers, defenders and protectors of good usage in Standard Written English. Let me now begin with a respectful salute to the great benefactors to Chemist's English. There are first of all those scientists (the name of L. F. Fieser comes most readily to mind, but some members of the distinguished tribe also reside in Australia) who by their passion for lucid prose have made others emulate them. Then, there are the great scholar-editors, among whom my special hero is R. S. Cahn. Men such as he have set standards for other editors to follow, and indeed the editors of high-standard journals and publishing houses form a team of beneficent Maxwell demons, allowing only expressions of high negentropy to pass into the language.

The status a good writer or editor has gained allows him, occasionally, to apply to a word a definition more restrictive than that used by the dictionaries. Now, the narrowing of a definition represents to the writer a gain in clarity, similar to the gain that the spectroscopist achieves by higher resolution. Let me explain by giving the Chemist's English definitions of "partly" and "partially".

In Standard Written English the two words are now full synonyms. It is a common bad habit, in writing, to use a longer word in preference to a shorter synonym, and thus partially is at the moment predominant, even though the letters *ial* are just "noise". In good Chemist's English, though, a distinction is made between the two words. As R. S. Cahn explains in the *Handbook for Chemical Society Authors* a "partially dehydrogenated product" is an intermediate compound in the dehydrogenation process (e. g. tetralin from decalin), a "partly dehydrogenated product" is one that contains unchanged starting material. *The Handbook for Authors of*

Papers in the Journals of the American Chemical Society concurs with these definitions, and helpfully lists the antonyms: "completely" is the opposite of "partially", "wholly" that of "partly".

Another word whose definition should be narrowed in Chemist's English ist that simplest of conjunctions, "when". It is inadvisable to use this as a synonym of "whereupon", as in "Barium chloride was added to sodium sulfate when a white precipitate formed".

Confronted with this type of sentence, I generally cross out the "when" and substitute for it, quite simply, a semicolon; this expresses the author's intentions more simply and more elegantly. But, according to the dictionaries, I have no right to interfere. *Whereupon* as a synonym of *when* is listed in all of them, although not, of course, as the primary meaning. The Shorter English Dictionary, without going into detail, gives 1803 as the date of the first appearance of a whereupon-like *when*, but according to the giant Oxford English Dictionary *when* has been used in very similar sense throughout the history of the language.

However, in the chemist's world the sequence of events has to be defined more precisely than in everyday life. Let us assume we start a timing device before the occurrence of two events A and B, and determine that they happen at the times t_A and t_B. Let us now consider the sentence "A happened when B happened" (or its equivalent "When B happened, A happened" — note that in each sentence the "when" clause encloses the B event). We shall substitute for A and B twice:

> We abandoned our work on examplamine when we found it had already been synthetized by Gorbushov. (1)

> A bromine solution was added when the tetrabromide precipi-
> tated. (2)

In sentence (1) we clearly have $t_B < t_A$ and in (2) just as clearly $t_A < t_B$. Now, surely in chemistry we cannot allow *when* to express two totally contradictory concepts. We can only admit one meaning, and that of sentence (1) has better credentials — the best of these is the fact that *when* in the sense of sentence (2) strikes us as rather stuffy in Standard Written English, and is almost unheard in Spoken English. (If the part of the sentence containing event A begins with *No sooner,* then t_A does precede t_B; I shall glibly argue that *no sooner* "reverses the sign" of t_A.)

Here endeth the most profound investigation into near-simultaneity since a certain paper appeared in *Ann. Phys.* in 1905. You may perhaps think that I have made too much of a fuss over the precise use of *when*; the word should be left to express a general relation $t_A \approx t_B$ and the reader should make up his mind about the time sequence from the context. Let me quote two sentences where this is not possible:

Heating was discontinued when the reaction subsided.   (3)

Acetyl chloride was added when the temperature reached 80°.   (4)

Well, has the reaction of sentence (3) been pushed to completion or hasn't it? And was the outcome of the event described in (4) an acetyl derivative or an explosion?

5   To Reflux or not to Reflux?

I had intended to make this the first series of articles on the English language in which Shakespeare is never quoted. But that was before I found, in the *Handbook for Chemical Society Authors,* the sentence:

> "Use of nouns as adjectives should be kept to a minimum; their use as verbs, *e. g.* 'to complex', is indefensible."

This made me think that perhaps it was just as well that Drs Cahn and Cross were not available to edit the First Folio. For if turning nouns into verbs is indefensible, then Shakespeare is the greatest offender in the language.

Open any of your Shakespeare volumes at random; on almost every page you will find an instance of noun-to-verb conversion. And the line that contains it will nearly always be the most vivid line of the page. The defeated Antony is deserted by the followers that formerly *spaniell'd* at his heels; Cleopatra worries about being *windowed* (exhibited) in Rome; King Duncan's presumed assassins are said to be *badg'd* in his blood. Shakespeare *prided* himself on his ability to use this verbal device; in fact he *gloried* in it; he *showered* the language with new words *coined* with a felicity that *beggars* description. And noun-to-verb conversion is still as common in our day as it was in the Elizabethan age; the chemist, with a sigh of relief, will note that no one can stop him from *centrifuging* his suspensions, and then *pipetting* off the supernatant.

Can we completely disregard the Chemical Society's prohibition, then? Hardly; it may not be literally correct, but it contains a core of good sense. Noun-to-verb conversion has lost the thrill of novelty it had for the Elizabethans; words no longer spaniel at our heels as they did at Shakespeare's. Moreover, the device, in our day, has been overused by headline writers in such manner that the newly coined verbs have an under-

tone of brusque and coarse action – on every front page some thug *guns,*
knifes, or *bombs* the hapless victim he has *lined* up.

Where does that leave the peaceful chemist? Everyone must exercise his
own good taste, but some tentative ground rules can be given:

a) Do not coin a new verb when all you gain thereby is the omission of a
simple word or two. "The compound complexes with" is no great gain
over "The compound forms complexes with".

b) Do not transform a noun into a verb where the language already
possesses a verb derived from the noun's stem. I think it was Professor
Guggenheim who pointed out that there had been no need for the
creation of *to bond;* we could with equal precision have said *to bind.*
Alas, this horse has bolted – to *bond* is part of the Chemist's English
now. But we can stop the swinish verb *to destruct* from competing with
to destroy in civilized (if that is the right word) language.

c) From nouns derived from Greek, Latin or French, verbs can often
legitimately be formed by suffixes such as *-ize, -ate,* or *-ify.* Wherever
such verb-formation is possible, do not use direct conversion. One horse
that has bolted from this stable is *to program(me).* It would only create
gales of laughter if some reformer were to advocate the use of *to*
prograph, and yet the scientific community has formed the correct verb
from the noun *chromatogram.*

d) A very strong prohibition: thou shalt not turn a cluster of nouns into a
single verb. *To bond,* as we have seen, is now correct; *to hydrogen bond*
is not.

e) And no proper names, please: let us not Reformatsky the language. Any
attempt to do so ought to be – ahem! – burked or boycotted.

How tentative these rules are, and how often they are disregarded,
becomes obvious when individual words are considered. Established usage
can overrule even the most sensible regulation. I shall discuss a few
noun-verbs that are at present in dispute.

To reflux seems to me now firmly established in Chemist's English. Its
use may be held to offend against rule (a), but the point is debatable. It is
strictly a chemist's word and unlikely ever to enter everyday English. I see
no objection to its use in the small print of experimental sections.

To thermostat is a vogue word at the moment. I think, though, that
authors feel slightly guilty about its use, for they nervously keep doubling
the final *t* to form the past participle *thermostatted* – to proclaim, no
doubt, their awareness that the word is not derived from *to state.* Now final
t is only doubled in stressed syllables. Hence, thermo*statted* is wrong and

thermo*stated* looks stupid; so why persevere with the word? Surely, "kept at $50 \pm 0.1\,°C$" is just as efficient a way to describe your procedure as "thermostated at $50 \pm 0.1\,°C$". Your learned audience will be able to guess that you achieved this constancy of temperature by means of a mechanical device.

In his thermostat(t)ed bath the chemist is likely to immerse a jacket(t)ed vessel. *To jacket* seems to me a justifiable verb-formation which does not offend against any of the rules given. But please do not double the final *t*. O well-dressed vessel, that may, at the same time and with perfect legitimacy, be vacuum-jacketed and silver-coated!

6 Amazing Revelations: English Scientists Secretly Practise German Vice!

A few more such Chapter titles, and this book can be offered for sale in supermarkets. The vice I am thinking of is the one of assembling a sausage-string of nouns and running them together to form a supernoun. This dreaded aberration is known as *Hauptwortkombinationenzusammen-setzungsbedürfnis.* The right-minded citizen averts his eye in horror from such multiple coupling.

Or does he? A literal translation of the above italicized abomination would be "noun agglomerate assembly mania", and is this expression really so remote from what passes for English in chemical journals? The awful truth strikes home that we, too, have been perverted; the only difference is that we shamefacedly try to hide our excesses by leaving spaces between the assembled nouns.

And worse revelations are to come. The Germans, at least, admit only nouns into their long sausage-words, whereas we, to lengthen the chain still further, cheerfully mix in a few adjectives, as in *proton magnetic resonance spectroscopy literature survey* or *cyclic ligand planar nitrogen array.*

These are shocking disclosures indeed, but let us overcome our panic and analyse the problem with scientific calm. We are dealing with a number of words $A_1, A_2, \cdots A_n$ which have been assembled into something that looks like a polypeptide chain:

$$A_1-A_2- \cdots -A_{m-1}-A_m- \cdots -A_{n-1}-A_n$$

This chain, according to right-thinking stylists, is of undesirable length. To break it down into manageable components, as any protein textbook will tell us, we have to study the nature of the bond between A_{m-1} and A_m. Analysis promptly reveals that there are two main types of bonding. In one, A_{m-1} is simply the genitive or possessive of A_m, and we have

$$\text{``}A_{m-1}-A_m\text{''} \quad \text{is identical with} \quad \text{``}A_m \text{ of } A_{m-1}\text{''}$$

An example of this type of bond is *literature survey.*

The other type of bond is represented by *protein textbook, trial run* or *dehydrogenation experiment.* Unlike the "genitive bond" above, this "adjectival bond" is rather difficult to express in different terms; as a general rule we can only say "A_{m-1}–A_m" is identical with "A_m having to do with A_{m-1}". Not that, in practice, we are likely to use the clumsy term *having to do with.* We say, rather awkwardly, "a textbook *on* proteins" and "an experiment *in* dehydrogenation". As for *trial run,* the bond between these nouns is almost impossible to break; the best we can do is to replace the noun-used-as-adjective *trial* with a "true" adjective such as *exploratory.*

From this analysis, two conclusions emerge. The first is that, far from being an out-and-out abomination, our A_{m-1}–A_m bond is a great asset to the language; it manages very subtly to express a relationship between two nouns which would otherwise be very difficult to be put into words. Scientific English would collapse without it, and even in spoken everyday English we have come to depend on it. Consider the sentence: "On my *television set,* I watched the *vice squad* investigate a *call girl racket*". Transcription, anyone?

The second conclusion imposes itself with equal force. An overlong string of nouns is bad English for the following reason. A noun-as-adjective or noun-as-genitive does not release its full meaning until we know the word it refers to. Thus, we do not know what A_1 is doing in the sentence until we have read A_2, and if that is followed by an A_3 noun we need to know this to appreciate A_2 and hence A_1. By the time we arrive at A_n we have had to memorize A_{n-1} nouns, and we are thoroughly infuriated with the author. (And, like as not, A_m will occur at the bottom of a right-hand page, so that the whole mental juggling act has to survive the turning-over process.)

Is there, then, a rule of thumb that allows us to determine how long our chain of nouns may be? Unfortunately not, but a few observations can be made. Suppose we want to make a statement about two polycyclic organic compounds that are similar but differ in the way two rings are fused together. If we suffer from the German vice, we might come up with:

> Results are not strictly comparable because of *ring junction carbon environment differences.*   (1)

Here we have constructed a "peptide" chain of the order A_5. Let us now "hydrolyse" this and investigate the products by what shall henceforth be known as "multiple-layer typography". We have:

   ... differences in the *ring junction carbon environment*   (2)

   ... differences in the environment of the *ring junction carbons*   (3)

   ... differences in the environment of the carbons at the *ring junction*   (4)

   ... differences in the environment of the carbons at the junction of the rings   (5)

   ... *environment differences* of the *ring junction carbons*   (6)

   ... *environment differences* of the carbons at the *ring junction*   (7)

Some other hydrolyses, such as $A_3A_4A_5 + A_1A_2$, are plainly not worth considering.

Of the versions given above, we can immediately rule out (1) as being totally obnoxious to the reader. Version (2) is the sort of thing that appears in the chemical literature but shouldn't. Versions (7) and especially (6) strike us as bad English; it is clear that noun chains should not follow one another too closely. Version (5), the product of total hydrolysis, arouses the reader's impatience by its fussiness.

We are thus left with versions (3) and (4). All textbooks on English usage would recommend the A_1A_2 version (4), but it must be admitted that the $A_1A_2A_3$ version (3) has its place in scientific writing, especially if the bonds in this chain cannot readily be split.

But there are two important exceptions to what I have just said. First, some chemical names are written as several words but the chemist reads them as one noun; thus I would class *lithium aluminium hydride reduction* not as an A_4 but as an entirely acceptable A_2 chain. *Cathode ray oscilloscope* and *nuclear magnetic resonance* also tend to acquire the status of a single noun (i. e. as soon as A_1 is pronounced, we know that A_2 and A_3 will follow) and certain cautious liberties can be taken by authors.

The second exception is this. If a noun is very long or demands great concentration from the reader, it is not suitable for incorporation into a chain. Thus we must write "reduction of 10-oxabicyclo[4,3,1]decan-2-one" rather than use an A_1A_2 version.

I could now close the *Hauptwortkombinationenzusammensetzungsbedürfnisdiskussion,* but before I do I'll let you into a secret. The greatest accumulators of nouns are neither the Germans nor the English but the Chinese. I hope this news fills you with relief; no vice ever seems completely abominable when you learn vast multitudes practise it too.

7 Of Nuts, Muttons and Shotguns

The preceding chapter dealt with noun chains. Or should I have written noun-chains? Clearly, the next topic we must explore is the use of the hyphen in written English. It is a rather complex topic and to do justice to it I'll have to divide my discussion into subsections. I'll begin by asking the profound question: when is a hyphen not a hyphen?

7.1 The Hyphen's Big Brothers

Print is easier to read than typescript. One of the reasons is that it does not have to rely on ambiguous symbols. Whenever your typist strikes the hyphen key, the resultant signal on the page may mean four different things, which the printer (prompted by the editor) clearly distinguishes.

(I) The *hyphen* of common-or-garden variety. This is a horizontal dash about as wide as the letter f.

(II) The *en dash,* called *nut* in the printing trade. Its name comes from the fact that it is about as wide as lower-case n. Whereas the use of the hyphen between two words indicates the merging of two concepts into one (an end-point of a titration is neither an end nor a point but a different thing altogether), the nut links together two concepts which retain their separate identities (a carbon–sulfur bond is a bond between two distinct atoms). Thus a careful editor will use the hyphen in a "blue-green solution" (*one* intermediate colour) but the nut in "blue–green dichroism". He will know that the symbol in "Lennard-Jones potential" (one man with a hyphenated name doing all the work by himself) is different from that in "Diels–Alder reaction". Apart from these subtle uses of the nut, the symbol is seen frequently in chemistry texts to indicate a range of numerical values (65–75% yields at 250–350 °C) and bonds between atomic symbols (C–O vibrations).

(III) The *em rule* or *mutton,* having the width of capital M. This can be used—and very effectively—to give emphasis to a parenthetic component of a sentence. When the emphasis is not warranted—as in this instance—its use is an enormous nuisance and the muttons should be promptly slaughtered to give way to parentheses or commas.

(IV) The *minus sign* occupies a space just as wide as that of the mutton, but the horizontal dash does not fill the width entirely. In proofreading, watch for the slight gap between the dash and the two symbols it links. In preparing typescripts make sure your typist leaves a full space on each side of a minus sign, else it may be misinterpreted as a nut and very serious errors may result. Thus M−CO will be interpreted by the reader as "a metal bonded to a carbonyl" whereas M − CO means "a molecular ion from which carbon monoxide has split off". (A similar precaution should be taken with "equal to" signs which can very easily be mistaken for double bonds.)

7.2   Hyphens Between Two Nouns

Be it understood that in this subsection I speak of groupings of two nouns *only.* It often happens that two nouns which, standing alone, do not need a hyphen *(trial run)* may need one when they are joined by a third *(trial-run results).*

Most textbooks state that hyphenation is the intermediate stage in the formation of a compound noun. According to this theory, writers begin with, say, *wave length.* After a while, familiarity with the concept makes them wish to run the words together but they do not quite dare, so they write *wave-length.* Eventually this hyphen is considered unnecessary and the compound noun *wavelength* is formed.

This is a plausible theory but in most instances it can be proved to be wrong. Does anyone suppose *compound nouns* will eventually become *compound-nouns* and then *compoundnouns?* And yet compound nouns have been around in learned literature for longer than wavelengths.

To counter this false etiology I shall now propose the Schoenfeld theory of hyphenation, which may be enunciated as follows: *in the marriage of two nouns, the hyphen is the shotgun.* Let me explain by an example: in my last article I used the expression "sausage-words". Sausages and words are a rather ill matched pair, and to create a link between them in the reader's mind I instinctively inserted a hyphen. And this, I think, is how most of the two-noun hyphens arise; they link the unfamiliar and can be dispensed with as soon as the unfamiliarity disappears. The two nouns may then either coalesce *(bookcase)* or stay apart *(filter paper)*. Whether one or the other happens depends partly on frequency of usage and partly on whether the new compound noun would look attractive to the eye.

It follows that, in two-noun combinations, authors should use hyphens only where the reader would be in genuine difficulty without them; and learned Journals should not allow their house styles to petrify but should revise their lists of hyphenated nouns at least every decade. And nobody, in trying to settle a question of hyphenation, should place overmuch faith in a dictionary, least of all in the COD. (I have often thougt the Fowlers would have hyphenated their grand-mother if they could have got away with it.)

7.3 *Hyphens for Pronunciation's Sake*

These have really no place in scientific papers, which after all are not meant to be read aloud. Persons who read chemical journals will have acquired sufficient education not to stumble over the unhyphenated versions of *coordinate* and *coworker*. To say that a hyphen is needed to separate the doubled vowel in *isooctane* is an insult to readers of chemical papers. And by inserting (as the Chemical Society does) a hyphen in *ethylenediaminetetra-acetate* we split the word at a point where logic demands it stay unsplit.

7.4 *Hyphens in Three-Word Groupings*

Here the situation changes radically. As we have just seen, a hyphen in *word groupings* is unnecessary; but as soon as these two nouns are prefixed by *three,* we need the hyphen to distinguish between *three-word groupings* and *three word-groupings.* Such ambiguities occur quite frequently in chemistry texts; for instance, in *complex ion mechanism* we cannot, without a hyphen, decide whether the ion or the mechanism is complex. In such instances, my advice to authors is to make use of the hyphen even if the chance of being misunderstood is small, as in *near-ultraviolet spectrum.* A good way to test the usefulness of a hyphen is to ask oneself the question: If the first word occurred at the end of a page, would a hyphen protect the reader from stumbling?

7.5   Hyphens After Adverbs

To be avoided. The suffix *-ly* forms a direct link with the word that follows, and so the hyphen is unnecessary.

7.6   Hyphen Clusters

We say, applying the end-of-the-page test, that a compound is *water-soluble.* No problem there; but is a compound *ethyl-acetate soluble, ethyl acetate-soluble,* or *ethyl-acetate-soluble?*

The first of these versions makes little sense and in fact I have never found it in a manuscript. The second version, to me, makes just as little sense but is commonly found in submitted manuscripts, and indeed is used by some publishing houses. (*Time* magazine replaces this kind of hyphen with a nut, which I think only makes things worse.) The third version seems to me the most logical, but most authors recoil from it in horror. Generally we settle matters by rewriting as "soluble in ethyl acetate" or we agree to leave out the hyphens altogether.

Why Australian authors should be so horrified by hyphen clusters I don't quite know. Used in moderation, they are a vivid, let's-try-something-new stylistic device, well suited to speeding up a sluggish sentence. I do admit that hyphen clusters should not be used in text which will be printed in narrow columns. Otherwise your let's-defy-tradition hyphens may well be diluted by end-of-the-line hyphens.

8 Tetravalency of Carbon Disproved!

You never expected to see *that* headline, did you? Alas, this chapter will not be as sensational as its title promises. Its aim is simply to discuss a modest recommendation by the IUPAC Commission on the Nomenclature of Inorganic Chemistry (embodied in its Rule 7.1). The Commission suggests that prefixes should be chosen so as to be consistent with the nouns to which they are joined. Now *valency* is a noun derived from Latin, and *tetra* is a Greek prefix; consistency demands that we replace Greek with Latin and say *quadrivalency*.

Sound scholarship, of course, is on the Commission's side. I dutifully pass their recommendation on. For the convenience of my readers, I shall list the numerical prefixes. The Latin ones are uni, bi, tri (or ter), quadri, quinque, sexi, septi, octa, ... multi. They are to be used with such Latin-derived words as -valent, -dentate and -molecular. Greek prefixes are mono, di, tri, tetra, penta, hexa, hepta, octa, ... poly. They go with endings such as -cyclic and -atomic.

At this stage the reader will experience his first misgivings. Do we really serve the interests of consistency by using different sets of prefixes for -molecular and -atomic?

Before I deal with these matters, let me first explain that the Commission's recommendations do *not* apply to chemical nomenclature (i.e. the naming of chemical compounds) itself. Here the Greek prefixes *always* prevail; thus we say *tetra*sodium pyrophosphate regardless of derivation.*

---

* Besides the "simple" multiplying prefixes di, tri, etc., there exists in chemical nomenclature a set of "compound" prefixes; bis, tris, tetrakis, pentakis, etc. In original meaning they resemble *twice* or *thrice*. They are used for multiplying complex expressions; thus bis(chloromethyl)amine is $(ClCH_2)_2NH$ whereas dichloromethylamine is Cl_2CHNH_2.

What with *tetra*sodium being mandatory while *tetra*valent is frowned upon, the Commission's plea for consistency is in serious trouble and, as the sensible scholars they are, they admit as much. Nevertheless their recommendation should be treated with respect. Few of us can lay claim to a classical education but most can guess accurately whether a word is of Greek or Latin origin and a correct choice of prefixes is the kind of verbal fastidiousness that gains respect from one's readers. We may, for instance, refer to a man who speaks several languages as *multilingual* or *polyglot,* but we would expose ourselves to much derision if we interchanged the prefixes.

So far so good. But while the Commission may win some battles in the fight against verbal untidiness (and indeed a survey of the literature shows that they have gained some territory lately) they cannot hope to win the war. To begin with, entrenched custom is hard to dislodge – we have all become attached to our *tetravalent* carbons and *monomolecular* layers. And then, the predominant language in chemical literature is English, and English is just about the most joyously untidy language there is.

Do you find this latter statement offensive? Let me remind you (pausing only to point out that *remind* was coined in the 17th century by joining a Latin prefix to an Old English noun) that present-day English was formed from three main reservoirs: 1. the Old English (OE of the dictionaries; itself an Anglo–Saxon–Scandinavian–Celtic mixture) as she was y-spoke before the Normans arrived; 2. the Old French (OF) imported by the invaders; and 3. the Latin and Greek words (some of these obtained by way of Modern French) added to the language after the Dark Ages ended. Each of these groups arrived with its own set of prefixes and suffixes (except that there was some overlap between 2. and 3.) and had the English had a passion for verbal tidiness they would have kept them apart. This, however, they were *un*able (OE + OF) to do, and the language became mongrel*ized* (OE + Greek via French).

This is not to say that the various sets of prefixes and suffixes can be sprayed about indiscriminately. The careful writer will fit an OE prefix, where one exists, to an OE noun. I still remember how, emerging after World War II from a Japanese internment camp, I was struck by the ugliness of the wartime word *de*louse. We have by now become used to *re*write and even *re*work but, to our credit, find words such as *re*say quite unacceptable. However, I have just read a novel by Anthony Burgess, no less, in which he uses the word re-met. And the intermingling of prefixes continues; let us admit that we rather enjoy referring to the splitting of our e.s.r. signals as *hyper*fine (Greek + OF).

Is there a rule of thumb which will allow the careful writer to choose the prefix or suffix appropriate to his word stem? I urge my readers to study the admirable article entitled *in-* and *un-* in Fowler's *Modern English Usage.* Its main conclusion is that, only where a word retains a Latin *appearance,* a Latin prefix may be preferable. Thus we have *un*able but *in*ability, *un*equal but *in*equality. But usage is capricious; if you are *un*certain about the appropriate*ness* of a prefix or suffix, resolve this *in*certitude by consulting a dictionary. I recollect that it took several tactful hints from my colleagues in the Editorial Service before I became aware that I had chosen an incorrect prefix for the word which usage spells as *un*obtrusive.

Untidy and capricious the English language may be, but it also has a matchless genius for putting its abundant riches to subtle use. By joining two apparently identical prefixes in turn to the same word stem, intricate shadings of meaning can be achieved. Thus a *sur*charge may be annoying yet legitimate, but *over*charging is illegitimate. If we say "Professor Winterbottom travelled from Conference to Symposium with the zest of an inebriate sailor lurching from grog-house to grog-house" then the comparison may not necessarily be *unapt* (i. e. a similarity may exist) but, applied to a venerable senior scientist, it is certainly *inept* (i. e. foolishly tactless).

I have a habit of thinking of languages as women (and well I may, having had love affairs with several of them). French, for instance, is a highly strung girl just out of convent school, forever nervously brushing foreign matter out of her skirt. English, now, is a cheerfully promiscuous wench, buxom and attractive; gaily bestowing her favours on whomever she fancies but always her own mistress; following no fashions but those that suit her. And if the dear girl decides to adorn herself with *tetravalency* then let her; it would look preposterous on anyone else but on her it seems right somehow.

9 This Chapter Explains . . .

When, a good many years ago, I became an editor, the prevailing opinion among my colleagues, and among reviewers, was that expressions like "Table 2 lists . . ." or "Figure 3 displays . . ." should at best be tolerated but certainly not encouraged. Tables and Figures, it was argued, are inanimate things; they list not, neither do they display. To ascribe these human activities to them was an unwarranted anthropomorphism; the authors of the offending phrases should if possible be persuaded to change them to the logically correct "Data are listed in Table 2" and "Results are displayed . . .".

I gained, in those days, the reputation of being an iconoclastic revolutionary by arguing the opposite. The ideal sentence in scientific writing, I said, was not necessarily the one that stood up best to rigorous semantic scrutiny but the one that conveyed information vigorously and unambiguously. By changing "the Table lists . . ." to "in the Table are listed" we would achieve no gain in clarity, but would only cause a formerly healthy sentence to limp along in the passive voice.

Time has passed and I have become a bit more conservative about defying the laws of semantics. But in the above matter I am certain I was right, and indeed objections to "the Figure displays" have disappeared over the years. Anthropomorphization of inanimate objects is an important human thought process; we cope with things better by giving them a human dimension. Thus we say "the car *drove* past" rather than *"was driven"*. We say "the ship *sailed* past" and may carry anthropomorphism to the point where we refer to the ship as *she*.

The chemist, of course, is so involved with his compounds and his instruments, with his data and his theories, that he is particularly prone to see them in human terms. In lab slang, we may say "I had a hell of a time *persuading* that ketone to crystallize" and in serious scientific writing we report, very vividly and quite properly, that a compound *resisted* oxidation and *accepted* two hydrogens under pressure. There is, come to think of it, a certain anthropomorphic quality about even such a neutral-sounding expression as "the compound *reacts*" (the word *to act* implies conscious participation) and we show we are aware of it by saying "it reacts *readily*".

All these expressions are very good Chemist's English and help to make scientific writing the vivid and exciting thing it ought to be. A note of caution must, however, be sounded. All good things can be carried to ridiculous excess. We have no right to say "the gas chromatogram *proclaims* the purity of the compound". If we see our chromatogram in human guise, we see it as a calmly logical being, not as an emotional one; hence *proclaims* is out but we can use *confirms* or *establishes*.

Likewise, it is bad style to say that "Table 1 argues" or "Figure 5 disproves". Tables and Figures, in their dispassionate way, *list, display* and *show information;* they don't get mixed up in the author's arguments. If some vivid word like *argues* is called for, then let us say "the results of Table 2 *argue*", "the shape of curve *C* in Figure 5 *militates against*" and so on. Data, results, curve shapes, rates of increase and so on can far more readily be imagined as participants in an argument than these pallid servants, Tables and Figures. Thus if there is any arguing, militating, agreeing and disproving to be done, it should be left to members of the former group. Or better still, the authors should come onstage and bravely say "On the basis of the data in Figure 3 *we* argue . . ."

(Incidentally, two authors recently have mixed up the word *militate* with *mitigate*. Please be careful. I know it's only a slip of the pen, and any Journal editor would put this sort of thing right by a flick of his wrist, without taking his mind off the scientific content of your sentence. But, during the internal refereeing process, you might just strike a sardonic fellow.)

At this stage my readers may be asking themselves what the point of my article is. The type of writing I recommend ("Figure 2 shows") is that practised by most authors anyway. Why preach at such length to the converted?

The fact is that I am trying to summon up courage to write about the extremely difficult problem of the *detached participle*. This has given me a great deal of worry; I cannot find a satisfactory resolution of the difficulties in the textbooks and to deal with the problem I'll probably have to invent an impromptu grammar of my own; the resulting explanation will stretch over three articles. Now, to clear some of the difficulties out of the road, I needed first to gain your consensus to the statement that anthropomorphization of objects and concepts, within sensible limits, is permitted in scientific writing. Let me explain.

"Detached participles" are participles (i. e. verbs turned into adjectives) which do not have, as all adjectives should, a noun to which they belong. Thus the italicized words are "detached" in "The compound was difficult to crystallize, *resulting* in considerable loss of material"; "The ketone was eventually reduced *using* the Wolff–Kishner method"; "The compound turned yellow, *suggesting* that autoxidation had taken place".

The sentences above are very common and also, to the purist, very wrong. *Resulting* and *suggesting* should belong to a noun but obviously do not. About *using* I have written before; if it is considered a participle it is incorrectly employed in the sentence above, but this situation may change if influential authors and editors accord it the status of a preposition.

It is at this point that the study of the detached participle joins up with our earlier disquisition into the legitimacy of anthropomorphic terms. In the sentence "the compound turned yellow, suggesting . . ." there is no satisfactory way in which we can link the noun *compound* to the participle *suggesting*. Compounds, as we have agreed, may *accept* hydrogen atoms or *expel* azo groups; fair enough, this is traffic between like and like. But they don't go around *suggesting* ideas to authors; this is anthropomorphy gone wrong.

But now let us change the sentence to: "a yellow color change occurred, suggesting that autoxidation had taken place." Immediately the sentence looks less offensive. A color change, it might be argued, takes place in the eye of the beholder; and the eye may well "suggest" ideas to the mind.

We can also come to terms with the sentence: "The method of Smith and Jones, *using* sodium hydroxide, resulted in 50% yield." A method, we may well say, can *use* a compound. Again, I have no objection to "The method of Smith and Jones was used, *resulting* in 50% yield". The sentence is not quite right, the author does not quite mean what he is saying, but at least he is not defying the rules of grammar aggressively. The editor is well advised to leave such a sentence alone. In my many years at the editing desk, for every ounce of English I've learned, I've learned a pound of tolerance.

10 The Painful Plight of the Pendent Participle – Preamble

If you have ever attempted to learn German, you will very soon have encountered the ultimate affront to linguistic logic. I refer to the German Subordinate Clause, wherein the subject from the verb by a vast stream of verbiage, which even a further sub-sub-clause include may, separated is. "Some clerks are responsible for this outrage" said my Viennese high-school teacher wrathfully (he was an ardent medievalist and much preferred Middle to New High German) "some wretched clerks at the court of the Habsburgs in the 16th century". I do not know whether the good *Herr Professor* was correct in his choice of villains, but the fact is that the use of this curious word order soon spread through the written and into the spoken language. In these days of ubiquitous radio and TV it has probably succeeded in dislodging the former word order, which Middle High German shared with English, from even the most isolated rural dialects.

This goes to prove that the human mind is capable of coping with sentence structures in which two words which should logically be close together are separated by wide gaps. Let us investigate the mental process by which this gap is bridged; it comprises three stages.

1. Recognition that a Subordinate Clause is just beginning. This is easy; the words that can initiate such a clause (relative pronouns, conjunctions) number only a few hundred and any six-year-old would know almost all of them.

2. An instruction to store the subject of the clause in the memory until the arrival of the verb signals the end of the clause.

3. Link-up of subject and verb, to make sense of the entire clause.

I am sure that in the heyday of cybernetics, when it was thought computers would soon do human translators out of their job, similar instructions would have been written into all programs for translation from the German.

Now, why this long digression into German syntax, in an article supposed to deal with participial constructions in English? The fact is that in each case the problem is similar: two words that should be closely linked have come adrift. But the mechanism whereby the listener's or reader's mind decodes the English participial sentence is more complex than that needed to unravel (apt word!) the German subordinate clause, and I thought it would help if we first worked our way through a simpler example.

If this talk of complexity makes you apprehensive, take comfort. English participial constructions are difficult to analyse, and it is easy to make mistakes in writing sentences of this type. But if any such mistake is uncovered in your manuscript, do not feel mortified; it happens to everybody. Shakespeare's characters slip up occasionally, and recently I found an example in a leading article, if you please, of that noble flagship of Fleet Street, the *Observer*. (The sentence was of the type: Believing *A,* it is possible to do *B*.)

The sentences that cause trouble in scientific English are of two kinds. In the first, the participle starts the sentence; i. e. it acts as "sentry" (the word is Fowler's) for a noun or pronoun that should be there but has been mislaid. An example of a wrong construction is:

> *Applying* the Woodward–Hoffman rules, the most likely struc-
> ture was found to be . . .   (1)

In the second type, the participle (generally *indicating* or *suggesting* or *resulting*) comes *after* a main sentence that seems complete in itself. Here is another wrong construction:

> The crystals were yellow, *indicating* that the double bond had
> shifted to a conjugated position.   (2)

I call such participles *advocates* because they plead with the reader to draw a conclusion. In the remainder of this chapter, and in the next one, I shall deal with "sentry" participles only; we shall return to "advocates", which require a different analysis, in chapter 12.

It is easy to see how the error in sentence (1) arose. The author (or authors) first formulated the thought in Standard Written English thus:

> *Applying* the Woodward–Hoffman rules, *I* (or *we*) found the
> most likely structure to be . . .   (3)

At the last moment he (or they) took fright at the pronoun of the first person, and attempted to translate the thought into impersonal Chemist's English. The result was sentence (1) in which the participle is left, as gammarians love to say, *dangling*. Actually, in our example it is not so much unattached as wrongly attached, for what sentence (1) says is that the structure (not the unmentioned author) applied the W−H rules.

The remedy in this case is simple. Overcome your fears of *I* and *we*, and transmute (1) into (3). *We* is by now generally accepted, and *I* is on the point of being accepted, in sentences where an opinion is expressed. Our problem at this stage looks trivial, and it seems that we can avoid the dangling participle by use of a three-stage program analogous to that discussed earlier, i. e. 1. recognize participle, 2. store in memory until the noun (or pronoun) it modifies arrives, 3. unite participle and noun to make sense of the "sentry" phrase.

But matters are more complicated than that. There is yet another way in which sentence (1) can be rescued, and in analysing it our three-stage program fails. We can write, quite correctly:

> *Applying* the Woodward−Hoffman rules leads to ... as the
> most likely structure.   (4)

In (4), as in (1), we have started the sentence with *applying*. In (4), as in (1), there exists no noun to which *applying* can be linked. And yet (4) is correct and (1) is not. What went wrong?

The answer − and of course my readers know this − is that *applying* in sentence (4) is not a participle (i. e. a verb turned adjective) at all but a gerund (i. e. a verb turned noun). Present participles and gerunds in English have the same form, and with this discovery our three-stage program lies in ruins. And worse is yet to come. Consider the following sentences, both correct:

> *Provided* with a mechanical stirrer, the *apparatus* func-
> tioned perfectly.   (5)

> *Provided* the air is excluded the yields are high.   (6)

In (5) *provided* is a participle but in (6) it is a conjunction. We are now faced with the full difficulty of the situation: a word which looks like a participle may in fact be a gerund, a conjunction, a preposition (e. g. *pending*) and other things more. It is easy to program a computer to recognize a German subordinate clause; all you have to do is to alert it to the occurrence of a few hundred words. It is well-nigh impossible to get a computer to recognize an English participle. You may teach it to recognize "participle-like words" by alerting it to the telltale endings *-ing, -ed, -nt, -lt,* etc., and making it check words with these suffixes against a thesaurus of verbs in its memory. When you have done this, and stored about a hundred more irregular past participles in the memory, the computer will recognize participle-like words with considerable efficiency; but it will not be able to discern whether they are actually participles.

But where the computer falters, the mind of, say, a ten-year-old carries on triumphantly. By now we know how he copes with the problem. He performs, all unconsciously, a multi-stage analysis:

1. a participle-like word (as just defined) is recognized;

2. it is stored in the memory and a search for the noun it may belong to is begun;

3. the sentence structure is scanned to see whether the word in fact performs the function of a participle;

4.1 if the word is not a participle the search is abandoned;

4.2 if the word is a participle, the search is continued till the noun is found;

5. participle and noun are linked.

Pressed for space, *I* had to oversimplify. There will be more about these matters in the next chapter. In the meantime, I leave you sentences (3) and (4) as protection against dangling participles.

11 Discussing the Sentry Participle, We ...

... recapitulate, first of all, its definition. A sentry participle (see Chapter 10) is one that occurs early in the sentence, ahead of the noun or pronoun it is linked to. If this noun or pronoun duly arrives (as the word *we* did in my opening sentence) all is well. If not, then the reader will stare at the sentence with the displeased air with which the chemist looks at a formula containing an element with an unsatisfied valency.

My earlier chapter on the subject contained some oversimplifications, which I shall now correct. When I wrote of "the noun to which the sentry participle is linked" I should have added that this noun (or pronoun) must always be the subject of the main clause. This is not a law imposed by schoolmasters on nature; the decoding process that we instinctively learn as children *conditions us* to expect a link between the sentry participle and the subject of the main clause. When this natural expectation is not respected, confusion results. Compare the correct sentence (1) with the dubious construction (2):

> *Sitting* in a London bus, *Kekulé* had an idea. (1)

> *Sitting* in a London bus, an idea struck *Kekulé*. (2)

At first glance, it would seem that (2) conveys the information just as efficiently as (1), but closer inspection reveals that the idea is described as the one doing the sitting. Now ideas may occasionally sprout wings, but they never come supplied with the anatomical feature that would allow them to take up a sedentary position in a London bus.

Sentences of type (2) are much beloved of pulp-paper novelists *(Sweeping into the drawing-room, a smile illumined Lady Cynthia's handsome features)*. They may seem a harmless lapse, but should be avoided because they can give rise to confusion. If you do not believe me, try replacing *an idea* in sentence (2) with *W. Perkin, Snr.* Here it would be impossible to tell which of the participants in the fictitious quarrel was sitting down, if it were not for our natural expectation of a link between participle and subject.

So much for poor battered Kekulé. The other oversimplification I have to correct is my definition of participles as verbs-turned-adjectives and gerunds as verbs-turned-nouns. They are that, of course, but they retain some properties that verbs have, and we should really think of them as hybrid forms. We have seen in Chapter 10 that in "applying the Woodward–Hoffman rules leads to . . ." *applying* is a gerund. If a gerund were simply a verb-turned-noun, *applying* here would be fully equivalent to *application.* But *applying,* owing to its hybrid nature, can still take a direct object (applying *the rules*) whereas *application* can only take an indirect one (application *of* the rules). Hence gerunds are a special subclass of nouns, subject to some rules from which other nouns are free. This will be very important later.

Armed with this knowledge, *we* are now almost ready to proceed to a refined analysis of the sentry participle. But we have to clear one last hurdle, that of the so-called *absolute construction.* I shall give an example:

> The constitution of the alcohol having been established, we
> turned our attention to that of the ketone. (3)

Having here looks very much like a sentry participle because it occurs early in the sentence. But it is not, for the noun to which it belongs, *constitution,* has occurred even earlier. Absolute constructions thus form no part of the subject we are investigating; I mention them only for completeness' sake and abandon their study with a sigh of relief.

From what has been said so far, and in Chapter 10, let us extract the two main facts. 1. If a participle-like word occurs early in a sentence, the reader expects it to belong to the subject of the main clause. 2. Sometimes this expectation is thwarted because the participle-like word turns out to be not a participle but a look-alike, such as a gerund.

Let us state our problem in terms that come naturally to chemists. There exists an *attractive force* between a participle-like word and the subject of the main clause. If the sentence is to have logical *stability,* this force must be either *satisfied* (by the discovery that the participle belongs to the subject) or *neutralized* (by the discovery that the participle-like word is not a true participle). Stated thus, our problem is familiar enough. Attempts to quantify the attractive force between reactive centres are common to modern physical chemistry; the literature is full of relative reactivities, Hammett coefficients and the like. Let us choose a coefficient of our own (*p,* the participle coefficient) to express the attraction between participle-like word and subject. Let 1 be the maximum value of *p* (the attraction generated by a "true" participle) and 0 the absence of any attraction. We shall now assign *p* values to various groups of participle-like words.

1. There is, first of all, the group of former participles whom usage has changed into prepositions. Examples are *pending, during, except, past, owing, notwithstanding, following, according, considering, concerning, touching, regarding, barring, including, respecting, seeing, following, granting, depending.* Those words have low *p* values (i. e. their occurrence raises no great expectations of a link between them and the subject). The first six in the list above could be assigned the *p* value 0; that of the others might vary from 0.1 to 0.3. Wherever the *p* value is not 0, there is some slight danger that use of the word may occasionally give rise to ridiculous double meanings. In such cases the word is meant as a preposition but can also be read as a participle: *"Touching* the high-voltage source, the *professor* said ...". *"Given* the enormous procreative power of the rabbit, the *investigators* decided to broaden the scope of their experiments."

The chemist need expect little trouble from these prepositions (the zoologist, as my last example shows, slightly more!) But what causes enormous trouble is the group of words which popular usage regards as prepositions but which are still considered true participles by dictionaries and established writers. We might say that these are words whose *p* value has just begun to drift from one. Among these are *assuming, based* (on) and the enormously important *using.* As matters stand at the moment, I can only advise my readers to use these words as true participles. But if they have strayed into your manuscript as prepositions, you need fear little criticism from referees.

2. Participle-like conjunctions. The only one that interests the chemist is *provided* (*except* as conjunction is archaic, *supposing* is slang). I have already mentioned it in Chapter 10. Shall we assign it *p = 0.3*?

3. Then there are some common phrases, such as *generally speaking, as mentioned above,* or *talking of . . .*, in which grammar is disregarded and the participle allowed to "dangle" by general consent. Occurrence of a participle in such a familiar locution causes its *p* value to drop to about 0.1. But not quite to 0; "broadly speaking, the lecturer said ..." could be understood by the mischievous to mean that the poor fellow had a notorious accent.

4. Now to the gerunds, these troublemakers. In many sentences the gerunds, in their capacity as nouns, are themselves the subject of the main

clause and thus could not possibly *dangle* (*Applying* the rules *leads* to . . . *Centrifuging* results in . . .). But there are other kinds of sentences in which the gerund follows a preposition ("after *cooling*") and in these the writer must be careful. Grammarians of the old school, such as C. T. Onions, stress the hybrid nature of the gerund and insist that a "sentry" gerund must refer to the subject of the main clause (i. e. $p = 1$). How much good sense there is in such an attitude can be seen from the hilarious misconstruction quoted by R. S. Cahn in the *Handbook for Chemical Society Authors:*

> After *standing* in the refrigerator overnight, *we* filtered off the acid.

Obviously the writers of this sentence did not know their Onions. But times are getting more permissive. Need we really object to a sentence such as

> After *centrifuging,* the solution was concentrated.   (4)

R. S. Cahn agrees with Onions in rejecting this construction. He points out that the sentence should read:

> After *being centrifuged,* the *solution* was concentrated   (5)

Here we have a proper link between participle-like word and subject. But Cahn, if I understand him correctly, would also permit:

> After centrifuging the suspension, the liquid was concentrated   (6)

This is no longer according to Onions, for the liquid is not doing any centrifuging. But the sentence shows what is being centrifuged clearly enough. One might say Cahn had a "$p = 0.8$" attitude towards gerunds. My own attitude might be described as $p = 0.6$. I would pass, with just a slight groan, sentence (4). The remedial sentence (5) is itself not quite perfect; one should really say "after *having been centrifuged* ..." But who wants to purchase correctness at the price of this kind of verbal tedium?

5. *Having* eliminated categories 1. to 4., *we* only have the category of "true" participles left. Here $p = 1$ rules rigidly. *Being* endowed with the nature of an adjective, the *participle* must belong to a noun or pronoun, and this should be the subject. *Bearing* this principle in mind, a *writer* should have no trouble with his sentry participles.

12   *The Case Against the Advocate*

Take heart, everybody; with this chapter we shall finish our survey of the tricky problem of the dangling participle. All that remains to be done is to rule on the legality of those participles which I, in Chapter 10, called "advocates". You may remember my example:

> The crystals were yellow, *indicating* that the double bond had
> shifted to a conjugated position.   (1)

"Advocates" such as *indicating, suggesting, resulting, proving* are thus participles that follow the main clause and begin a new clause in which, generally, some conclusion concerning the facts of the main clause is drawn. Let us have another example:

> The compound proved hard to crystallize, *resulting* in reduced
> yields.   (2)

Resulting here is a dangling participle; the only noun it could belong to, *compound,* will obviously have nothing to do with it. Sentences (1) and (2) thus offend against a grammatical rule and one would expect to see them vigorously attacked in the appropriate textbooks.

Strangely enough, the learned compendia, which are full of savage condemnations of the dangling "sentry" participle, have very little to say about the "advocates". This is all the more deplorable as the issue is by no means clear-cut; it is possible to defend advocate constructions on grammatical grounds. One could say that these constructions are correct sentences but for the omission of the words *a fact* just in front of the advocate participle:

> The compound proved hard to crystallize, *a fact* resulting in
> reduced yields.   (3)

Sentence (3), though it now reads rather awkwardly, has been cured of the dangling participle, and the same treatment could be applied to sentence (1) and any other advocate construction. Now, the omission of words the reader's or listener's imagination will immediately supply is a well known linguistic process; grammarians refer to it as *ellipsis.* In leaving out words [that] it does not need the English language is as efficient as any [other language]. [Do you] See what I mean?

Advocate sentences could thus be justified as "elliptical constructions". The scientific editor, in deciding whether such a justification has merit, receives little help from existing treatises on English usage. How does he make up his mind?

He takes note, above all, of what contemporary writers of good repute are doing. Advocate constructions, to the best of my knowledge, occur very rarely in modern novels – a good writer of fiction would reject such constructions as too ponderous. They do occur, but not too often, in the politico-literary magazines whose reputation attracts distinguished contributors. They occur more frequently in news magazines, even in those which pride themselves on their brisk and lucid prose. Yet even in these publications I have the impression that advocate constructions are tolerated rather than encouraged.

I can only express my own opinion about what conclusions a scientific editor should draw from this evidence. When I encounter an advocate sentence, my first concern is to see whether the participle must definitely be considered as dangling. In many cases a participle can be considered to be linked to a noun by the process of anthropomorphization about which I wrote in Chapter 9. If you look back at sentences (1) and (2) you will see that (1) makes much better reading (our imagination can cope with crystals *indicating* something, but a compound cannot do any *resulting*).

In many cases, thus, the objection against the advocate participle can be overruled. Sentence (2), for instance, would become acceptable if recast as follows:

> Crystallization proved difficult, *resulting* in reduced yields. (4)

But there are cases where this loophole is closed, the participle is clearly detached and the sentence has an awkward look. In such cases I often replace the comma of the original sentence with a semicolon and the participle *x*-ing with "this *x*-ed":

> The compound proved hard to crystallize; this resulted in low yields. (5)

I confess I make this change very hesitantly. No author likes to see his manuscript interfered with; is it worth annoying a valued contributor over a trifling point of grammar? I believe that the change from (2) to (5) is just sufficient improvement to commend itself to authors, and shall set out my reasons (if you disagree, please let me know). In the first place, I feel the semicolon is more in the spirit of an "advocate" sentence than the comma. The evidence has been presented in the first part of the sentence; the pleading is about to begin, and the semicolon represents the expectant hush in the courtroom. Second, the participle (an inactive verb form) of the original sentence has been replaced by an active verb, which breathes new life into the amended sentence. Against this, it must be admitted that the amended sentence contains the pallid word *this,* which is more noise than signal.

I hope I have gained a little sympathy for the editor's predicament, and for the change from (2) to (5). But the editor should not have been put into this predicament in the first place. If, on rereading the first draft of your paper or report, you come across a sentence that seems awkward or cumbersome, the remedy is not to fiddle with verb forms or punctuation until the amended version will just get past the reviewer, but to recast the sentence completely. Thus, far better than (2), and the fumbling rescue attempts (4) and (5), is the sentence:

> The compound proved hard to crystallize, hence yields were low. (6)

After the original version of my article on the "advocate" appeared, Dr W. S. Cohn, of Oak Ridge, Tennessee, reproached me for not pointing out the following alternative to sentence (5):

> The compound proved hard to crystallize, *which* resulted in lower yields.   (7)

Dr Cohn is of course right and I accept (7) as being the equal of (5) in effectiveness. Observe that here *which* is not attached to any noun but stands for *a fact which;* thus it relies for its acceptability not on logic but on usage. Shakespeare's contemporaries, the founding fathers of modern English, would have said *the which* instead of *which.* The which (to adopt their language for a moment) would have been best. 'Tis pity that our language should have become so enriched in its store of words since Shakespeare's day, yet so diminished in pungency.

To sum up: solutions (5) and (7) are acceptable means of dismissing the troublesome advocate from your sentence (2). But solution (6) is much to be preferred. In the Chemist's and anybody else's English, the simplest way is always the best.

13 That's the Way She Crumbles, Language-Wise

Three decades ago the -*wise* craze spread through all English-speaking populations. The suffix -*wise*, of course, was not a new coinage; it had long existed in the language in two accepted usages. The term X-wise could mean 1. "wise in the ways of X" (as in a *weatherwise old sailor)* or, more commonly, 2. "in the manner of X" (as in *dropwise, clockwise)*. To these accepted usages, another was now added: 3, "as far as X is concerned". Devotees of the new fad might thus say, for instance: "We haven't checked the results out yet, statistics-wise."

The fad is now dying out. In a way I am sorry; it had a brash cheerfulness about it and was at least shorter than some of the cumbersome phrases it replaced. But like most fads it died of overuse: also it became identified with a certain vulgar subculture. Soon it became an object of satire for writers such as Billy Wilder ("That's the way she crumbles, cookie-wise" says the hero of the movie *The Apartment*). There was also the *Punch* cartoon in which an owl enquires of another concerning a fledgling third: "How is he shaping up, wisewise?".

Into the Chemist's English the fad hardly ever penetrated. Only in one instance in my editorial career did I ever have to move my word-wise pen sidewise through a word to make a correction *wise*wise. What we do say in the Chemist's English is: "these results have not yet been *statistically evaluated*". The present article is an enquiry whether *statistically evaluated* (*chromatographically purified, spectrometrically analysed* etc.) is good English usage or whether it is not just a roundabout way of saying *statistics-wise*.

This question is causing me some concern, for it seems to me that the -*wise* and -*ally* phrases form a part of a new phenomenon which I call the Mongolization of the English language. (Mongol languages. I have been told, are characterized by the frequent use of suffixes and postpositions.) Let us assume that in a sentence containing the subject Z and the passive verb Y a noun X should form an indirect object to the verb. The normal English version would thus be "Z was Y-ed *by* (or *to, against, with* etc.) X". The "mongolized" sentence would run, however: "Z was X-ally Y-ed" or "Z was Y-ed X-wise". Authors of such expressions apparently shy away from the necessity of enclosing two nouns Z and X in one sentence. They take X, add to it an adverbial-type suffix (*-ally* and *-wise* by no means exhaust the list) and steer it hopefully in the general direction of the meaning.

Are they entitled to do so? One could argue the point. The dictionaries define X-ally as "pertaining to X" or "having to do with X". The expressions cited above do not fit comfortably into these definitions. What we are really trying to say is "in accordance with X" or "by an X method"; this is not quite the same thing. Occasionally, when we are trying to write a sentence in good Monglish, a meaning nearer to the dictionary definition appears briefly between the lines. We say "The course of the reaction was *energetically* determined" and for a moment the reader misreads this as "we vigorously made up our minds what the product should be". Or the sentence "He was *constitutionally* prevented from standing for a second term as President" carries undertones of "His first term gave him a nervous breakdown, poor chap".

But dictionaries, in such matters, are as unreliable as the referee at the ball game who is 50 meters behind the play. *Chromatographically purified* and kindred expressions are by now established usage; they are regularly used by the educated, and sooner or later the compilers of dictionaries will include "in accordance with X" among the definitions of X-ally. I am not attempting to turn the clock back; all I would like my readers to do is pause for a moment before writing *spectrometrically analysed* and ask themselves whether *analysed by spectrometry* might not read better in the context.

Let us also promise each other not to let the meaning of X-ally stray too far beyond "in accordance with X". When I read sentences such as "The discharge of sewage is *municipally prohibited*" I seem to hear the thunder of hooves swiftly approaching across the Kirghiz steppe.

So far I have only dealt with the "adverbial type" of Monglish. There is also an adjectival type of Monglish, closely related. Here a noun R is meant to be joined to another noun S by means of a preposition: R *to* S, R *about* S, R *against* S, and so on. Under the Mongol influence, we shrink from the use of a preposition and append to the noun an adjectival suffix, the more pompous-sounding the better: R-ic, R-(ic)al, R-iferous, R-ogenetic. Very often the resultant R-ic S phrase will be quite correct: *spectrometric* means "spectrum measuring" or "pertaining to the measurement of spectra", hence it is all right to talk of a *spectrometric determination*. Over the phrase "the *spectrometric behaviour* of examplitol", however, flutters the banner of Genghis Khan.

The most assiduous purveyors of this kind of Monglish are, fortunately, not chemists but social scientists. These good people, in their entirely justified desire to have their field of endeavour accepted as an exact science, hope to achieve their ends by generating as much jargon as possible. Hence they will mongolize "analysis of a situation" into "situational analysis". I have even seen a "response to a film" rendered as "filmic reponse". But let us not sneer too soon. One American journal of chemistry has published a survey of the "osmic literature".

Sometimes Monglish crops up in unexpected places. The abdication talk of Edward VIII has been justly praised as a sincere and direct speech. However, near the end, the departing monarch referred to himself as "bred in the constitutional tradition of my late father". What *constitutional tradition,* Highness? Surely, what you meant was a tradition of respect for the constitution. We know now that this particular expression was suggested by Winston Churchill; it is much the worst thing in the speech. Churchill was a great orator in his day, but at times he would reach for the grand phrase as a drunkard reaches for the bottle.

It must be admitted that Monglish has its uses; at times it allows you to generate gobbledegook when gobbledegook is expected of you. For instance, in census or tax forms, I always list my occupation as "scientific editor". Now I have no way of knowing whether I am a scientific editor; all I know is that I am an editor of scientific texts. But the phrase has a reassuring sound, and where would we all be if we inflicted an exact description of our jobs on a hapless tax clerk or census taker?

As an example of how fast the invaders can travel once they have spotted a breach in the Great Wall, I offer the use of "hopefully" in the sense of "it is to be hoped". Thirty years ago this was almost unknown in Australia; twenty years ago this would have been considered here an Americanism; now all the newspapers are full of it. I have already had occasion to praise, in these columns, Th. M. Bernstein's *The Careful Writer.* In a brilliant analysis, Bernstein shows that *hopefully* is the reverse image, as it were, of *regretfully* but not of *regrettably*. Regretfully, I note that this regrettable Monglish misuse of a suffix has become commonplace; however, the *New York Times,* the *Australian Journal of Chemistry* and, one hopes, other publications of high intellectual endeavour still hold out. The use of *sadly* as in "sadly, there are no research grants available" I denounce as an utter barbarism. The Monglish *hopefully* is at least trying to fill a vacuum, but the niche Monglish *sadly* tries to occupy is already inhabited by *regrettably* and *unfortunately*.

But let me return to my theme of the speed with which the language can be corrupted. Bernstein's book appeared in 1965; Sir Ernest Gowers's revision of Fowler's *Modern English Usage,* which was published in the same year, is still unaware of the misuse of *hopefully*. (Sir Ernest — no doubt out of disdain rather than ignorance — also ignores the *-wise* craze; Bernstein dissects it with hilarious brilliance.)* Both word-witted writers are silent, as far as I can see, on such constructions as *municipally prohibited* and *situational analysis,* nor have I found anything on the subject in the other authorities. Perhaps I am making a great fuss over very little. But we live in a beleaguered society, and any citizen who believes he has spotted an approaching threat has the duty to let the cry "The Mongols are coming" resound from the fortress towers.

---

* In a more recent and equally excellent book, *Dos, Don'ts and Maybes of English Usage* (Times Books, 1977), Bernstein *regretfully* and (I think) *regrettably* concedes defeat in the matter of the unfortunate usage of *hopefully*. I don't know whether Bernstein is speaking for the *New York Times* but he definitely does not speak for the *Australian Journal of Chemistry* nor for the top British newspapers, weeklies and magazines.

14 Now, from the Pen that Gave You Monglish, Comes Gerglish

Much time must necessarily pass between the writing of a chapter and its publication in a book, and it is just possible that the phrase will have fallen from favor when this appears in print. But at the time of writing, one cliché reigns supreme. Whenever the activities of an organization or government department are described, sooner or later space will be found for the following arrangement of words: "This is part of an ongoing process".

Now nobody should begrudge the public relations men their little ritual. The sentence suits their purposes admirably. Firstly, it conjures up an image of a department full of hustle and bustle. Second, by implying impermanence, it disarms criticism: "So you don't like our structure? But it is about to change! Onkeep your shirt, brother, things are still ongoing!"

So far so good. But why have the mass assemblers of stock phrases abandoned such useful standbys of yestermonth as *continuous, dynamic* and *evolving* for *ongoing?* How long has this been ongoing, anyway?

Well, to be precise, since 1882. In that year, according to the giant *Oxford English Dictionary, ongoing* in the sense of continuous was first seen in print, oddly enough in a dictionary, compiled by John Ogilvie, which paid special attention to scientific and technical terms. But Ogilvie, as a lexicographer, seems to have been ahead of his time. The word did not come into general use and is not listed in the medium-size English and American dictionaries printed around 1940. After the Second World War, it seems to have caught on (oncaught?) in the United States, for both *Webster's Third* and *Random House* list it readily and are profuse with quotations. But my copy of the *Concise Oxford,* printed in 1964, still ignores it entirely. And now suddenly it is smiling at us blandly from every press release.

Now it is bad enough having to put up with one uninvited guest, but what if he brings his family? Vogue words always travel in packs. Say two decades ago, if you spotted on a page the word *heuristic,* you could be sure that *epistemology* or *dichotomy* would not be far away. Nowadays, within five lines or so of *ongoing,* you are bound to spot *input* (attended by his cousins *throughput* and *output*) and *feedback.* All the new vogue words have one feature in common: they seem to be the offspring of a union between a verb and a preposition: break*through,* count*down,* fall*out,* *down*turn, *up*swing. Some of these words are most admirable inventions. *Throughput* may not sound very nice, but I could no longer manage without it. And *feedback* is so vivid a word that it has made a difficult philosophical concept accessible to the man in the street. My quarrel is not with any individual word (I suppose I could even learn to tolerate the super-fluous *ongoing*) but with the fact that, like mushrooms, they grow in clusters. And their continual use makes for stuffy and ponderous prose, reminiscent (for a reason I shall explain later) of German technological writing. My pet word for it is Germanized English, or Gerglish for short.

What follows is a sketchy and highly tentative explanation of how Gerglish words arise. It seems to me that a verb V can engage in two kinds of traffic with those words that the dictionary lists primarily as prepositions. The first kind reminds me of the covalent bond. The former preposition turns itself into a prefix p and is permanently joined to V to form a new verb pV which undergoes all the customary transformations: I pV, he pVs, he is pVing, etc. An example is *under + take = undertake,* hence *undertakes, undertaking.*

The other possible linkage between verb and "preposition" resembles the ionic bond. The two components engage in a union which profoundly alters their properties; when we say "the machine *breaks down*" neither *break* nor *down* retains its original meaning. The preposition changes into something which can be vaguely described as an adverb A, but V and A are not joined together. A always follows V, sometimes at a distance. I VA, he Vs A, I must V something something A. For example, I must *break* this article *down* into subsections.

German, like English, has its pV "molecules" and VA "ion pairs". The "ion pairs", however, have a curious property: they exist only when the verb V is in an active form. In all inactive forms, such as participles and gerunds, V and A "crystallize out" into the firmly bound form AV. Thus, "the investigation *goes on*" in German becomes "die Untersuchung *schreitet fort*" (literally, *strides forth*). "The investigation *going on*" is "die *fortschreitende* Untersuchung" (*forthstriding* or, if you will, *ongoing*). And the derived noun is the familiar *Fortschritt,* progress.

The noun formation, AV, is thus decreed in German by both grammar and custom, and the technical literature fairly bristles with such nouns. In

English, however, the "crystallization" process is not automatically permitted (there are no such nouns as *ongo* or *forthstride*) and when it occurs it may give rise to either *AV* (*in*put) or *VA* (feed*back*). The only technique of noun formation from *VA* pairs that is unconditionally sanctioned by English grammar is the use of the gerund *V*ing *A* (as in, the *breaking up* of the molecule) and I appeal to my readers to use this device to break up the grim monotony of Gerglish prose.

Two fascinating observations about English prose clamor to be made. The first concerns the coexistence of a *pV* verb with the word pair *VA* in the case where, formally, *p* = *A*. We have seen that a *VA* pair may, occasionally, "crystallize out" as the noun *VA,* so that both *pV* and *VA,* where *p* and *A* are outwardly identical, may occur in the language. (In German this cannot happen, because German only has the crystal form *AV,* which would be undistinguishable from *pV.*) This may give rise to much entertaining word play. If your company is *overtaken (pV)* by a *takeover,* you may be *upset* about the new *set-up.* Or it might cost my publishers a considerable *outlay* to alter their *layout.* I find such sentences not merely amusing; they inspire me with absolute awe for the subtleness and ductility of the communications system which is the English language. Take a phrase such as "He may not *last out* the year, but his work will *outlast* the century". In such a sentence, modern English stands revealed in all the glory of its resourcefulness.

The second observation is really a question. Why sometimes *AV,* sometimes *VA,* sometimes nothing at all? A fire breaks out, a pump breaks down – one event is an *out*break, the other a break*down.* Excessive use of the flash*back* in a film may produce a *back*lash of annoyance in the audience. I must confess I do not know why word formation is so capricious; I can only offer tentative explanations. *AV* words are formed by analogy with existing *pV* words (perhaps *outlast* led to *outcast*) or derived from the German (*throughput* may well have been inspired by *Durchsatz*). And of course, the appearance of one *AV* word leads to the formation of clusters of similar words (*income, outgo; input, output*) and to a temporary fad of *AV* creations.

The fad seems to have been at its height in the 19th century. Most of the *AV* words I have traced through the dictionaries are 100–200 years old (a very few were formed "ahead of their time" – Shakespeare bewept his *outcast* state in his sonnets). The *VA* word is, generally, a child of our century (*feedback,* first applied to electrical circuitry, is not quite 60 years old). The *AV* word was generally created for written communication, the

VA for speech. Its godfathers were the slangs of the film, jet propulsion and computer industries. And it found a powerful patron in the newspaper headlines: *cover-up, shutdown, drop-out.* Many of these words are so new that they still need the hyphen.

May I proceed to a *summing up* (not a *sumup* nor an *upsum*) of this article? *A V* and *VA* words are both firmly established in the language, but excessive *cluttering up* of your prose with such terms will mark you as a poor stylist. Too many *A V* words will cause your writing to seem bookish and ponderous, too many *VA* words will make it slangy. Our investigation has shown how muscular and lithe the English language can be. Don't make her flabby by feeding her such useless pap as *ongoing.* If you do, you will find that there is a continuous, dynamic and evolving process taking place in editorial offices which is known, not as crossout or outcross, but as crossing out.

15 The Chemist and the Capercailzie

The Chemist's English has borrowed a multitude of words from Standard English and given them special meanings; think only of *ring*, *chair* or *bed*. But the traffic has not all been one-sided; we have given Standard English such words as *throughput, catalyst, quantum jump* and *interface*. I must confess that I feel a bit uneasy whenever I note one of our beautiful specialist words in unhallowed mouths or pens. *Catalyst*, in a recent issue of the *London Observer*, has aquired the adjective cataly*stic*. *Quantum jump* is often quite felicitously used by journalists for a spontaneous and immediate change; just as often, however, it is misused to mean "a very large jump". As for *interface*, this has become a vogue expression among social scientists. But it seems to me that in their hands it remains a drab cliché, whereas to the chemist it conjures up a vivid image of a glittering sheet of refracted light stretching between the upper and nether layers in the separating funnel. But perhaps I am prejudiced.

Anyhow this article is all about interfaces; about what happens to words or letters when they diffuse from one culture or subculture into another. There is, for instance, the word *lecture*. It comes from the Latin "to read" and in French, very logically, it means "perusal". In English, however, it has perversely come to mean a spoken address.

15.1 A Synthesis Simply not Facile

Or take the word *facile*. In Latin *facilis* means basically "that which can be done"; it acquired the meaning of "that which can be done with little effort" and the antonym *difficilis*. These two words "diffused" through the Latin–French interface without alteration in meaning, and with only the formal change to *facile* and *difficile* (the noun of the latter is *difficulté*). When these two words migrated to England the second resurfaced in the form *difficult;* its meaning was still precisely that of French *difficile.* The word *facile,* on the contrary, survived in unchanged form but with subtly altered meaning. During the channel crossing it had somehow acquired a derogatory connotation, with undertones of slickness and shallowness. Thus a *facile victory* was hitting a man when he was down, a *facile solution* one which would temporarily appeal to everyone present but would not stand up to hard reasoning the morning after. The *Handbook for Chemical Society Authors* states firmly: "A synthesis may be simple, useful, ready, easy, etc.; it can hardly be facile."

But 25 years have passed since the *Handbook* was printed. What I am going to say may bring me some angry letters, but I am of the opinion that, in diffusing into the Chemist's English, *facile* has regained its old French–Latin meaning and is again a full antonym of *difficult.* Chemists have tried all the words so helpfully offered by the Society and found them inadequate in certain circumstances. When you say that yours is a *simple* synthesis, you imply that the man who prepared the compound before you by a different path was rather an idiot. *Useful,* with all respect to the *Handbook,* does not mean the same thing. A *ready* synthesis does not sound right and carries implications of mild reagents and room temperature conditions. An *easy* synthesis is one you can hand over to the lab boy. I am afraid facile syntheses and rearrangements are with us to stay. In my more zealous days I tried very hard to get rid of the word but all the would-be synonyms I came up with offered only facile solutions, in the old sense!

15.2   Of Ghosts and Pirates

Then there is *spectral.* Some very fastidious writers still avoid terms like *spectral properties;* to them *spectral* means exclusively "pertaining to a spectre", i. e. ghostly. But the chemist, though he may be haunted by gremlins, deals in his published work with spectra and not with spectres. The word *spectral* in the sense of "pertaining to a spectrum" has come to stay, and is sanctioned by the dictionaries.

The enchanting word *moiety* has all but disappeared from everyday English, but fortunately has found a good home in Chemist's English.

Moiety was, in its day, much beloved by writers of adventure stories, and for me it still conjures up visions of fiercely mustachioed and be-cutlassed pirates greedily dividing up a pile of doubloons. To the best of my knowledge, the man who introduced it into Chemist's English was R. R. Williams, the remarkable chemist who established the structure of thiamin while working for the Bell Telephone Company. Modern science administrators will think back with horror to the days when scientists were allowed to do what they liked: here was this man who should by rights have worked on insulating materials, and instead he just self-indulgently went ahead and rid the world of beri-beri. When using *moiety,* though, please note that, coming from French *moitié,* it means "half" or "almost half". It cannot be used for a small fragment of a molecule. And please also note the spelling.

15.3   -Ise? -Ize? -Or? -Our?

This brings me to the subject of spelling, which I have so far earnestly avoided in these articles. I have been asked to express an opinion on the endings *-ize* or *-ise* (as in crystal*ize*) and *-our* or *-or* (as in col*our*). And this will fit in well with this article, which deals with the diffusion of words from one culture to another.

There is a widely held belief that the endings *-ise* and *-our* are older (they are not) and more "educated". This last point can be disputed, and indeed has been passionately disputed by the leading authorities of this century. Partridge, for instance, says in *Usage and Abusage:* "to employ *-ise* is to flout etymology and logic". Fowler is just as firm in his support of *-ize* and recommends, in moderate terms, the dropping of the *u* in *-our.*

It all boils down to whether you like to buy your suffixes from the manufacturer (Greek and Latin) or the middleman (French). The Greek *-izein,* written with a zeta, became *-izare* in Late Latin, then diffused into French where it was eventually spelled *-iser.* The first use of an *-ize* word that the *Oxford English Dictionary* knows of occurred in 1297; the word was *baptize* and was spelt with a *z;* in another document dated 1300 the same word, however, has an *s.* The muddle has thus been with us for 700 years now. As for *-o(u)r,* the original Latin spelling was *-or;* words like *color* and *honor* became *couleur* and *honneur* in French. Anybody who has got as far as the third line of the *Canterbury Pilgrims* knows that the *-our* spelling was favo(u)red in Medieval English, but since then the vowel (then pronounced as *ow*) has lost its stress and most of the now unnecessary *u* letters have been shed: the word Chaucer used is now *liquor*. The band of *-our* words is now rather small (the Americans, of course, have done away with the *u* altogether) and those that remain lose, in modern usage, their *u* when joined to the suffixes *-ous, -ation* and *-ize: coloration, vaporize*. Some readers may want to vapourise me for saying so, but they can find it all in Fowler as revised by Gowers, p. 428.

In Australia we have the following grotesque situation. The daily press uses *-ise* and *-or*. The learned press (including the Journal I help edit) generally follows the *Concise Oxford Dictionary* and prints *-ize* and *-our*. The Government Public Service *Style Manual* first recommends the COD and then reverses itself and opts for *-ise* and *-our,* but, at the time of writing, this spelling has not commended itself to the Department of Labor.

Personally, I am an *-ize* man, and not only for reasons of etymology and logic. The publishing firm which dominates the market for English dictionaries in Australia and Britain, The Clarendon Press, is giving *z* as its first preference, and it is good policy to "get with the strength". I also believe that, now that English has become the international language of science, we ought cautiously to move towards better phonetics in spelling. And, in these days when linguistic processes are being reversed, and languages which were drifting apart are brought together again by radio and Telstar, ought we not to diminish the differences between British and American spelling?

The main argument in favo(u)r of *-ise* is that it avoids confusion with some two dozen words, such as *surprise* and *advertise,* that do not derive from *-izein* and that are legitimately spelled with *s*. But the Americans can cope with these words; why should they cause difficulties in other English-speaking countries?

15.4   The Remarkable History of the Letter z

While reading up on the history of *-ize* I was struck by the appearance of *z* in Late Latin *-izare*. The Latin my high school teachers had so unsuccessfully tried to din into me had not contained this letter. This made me curious and I am now able to tell you the life story of the letter *z*. Like most consonants in our present alphabet, z comes originally from North Semitic script; the symbol apparently represents the shape of a weapon whose name, *zayin,* began with the *z* sound. The Semitic alphabet was adapted by the Greeks; if your dictionary contains a table of alphabets you may note the vestigial resemblance between modern Hebrew *zayin* and Greek *zeta.* The Etruscans adapted the Greek alphabet to their language and the inhabitants of Latium in turn formed the Latin alphabet from Etruscan letters. This alphabet originally contained *z* as (like Greek zeta) the seventh letter. However, the sound it represented (*ds* between two vowels) vanished from the language, and around 300 B. C. the letter was dropped from the alphabet. The newly formed *g* became the seventh letter. Then followed the Roman conquest of Greece and the flirtation between the Latin and Greek cultures. The letter *z* was reinstated in Cicero's day to transcribe the zeta sound in Greek words; hence *-izare.* But in the course of history *z* had "missed promotion"; it now became the twenty-third letter.

Z survives vigorously in Italian, which is just as well, because a *pizza* in another spelling might not have seemed as tasty. In French, as the *d* disappeared from the original *ds* sound, the number of *z*'s diminished, and English conformed to French.

But *z* had an entertaining adventure when it crossed into Scotland. Scottish scribes had used a letter looking very much like written *z* to represent a "y" sound. From 1300 on this became confused with *z,* and of course the import of standard typefaces after the invention of printing accelerated the process. In time people came to pronounce words as they saw them written, and by now all Sassenachs, and alas many Scots, mispronounce the names of Menzies, Dalziel and Mackenzie. The only holdout is the *capercailzie.* The bird is commonly written with the *z* but mostly pronounced with the original faint *y* sound, to rhyme on "Och, wha' can ail ye?"

So, whenever you get annoyed with your editor for changing one of your suffixes to whatever his house rules dictate, consider the capercailzie. Think of this large grouse, hiding in the gorse from Caledonian weather and kilted hunters, and having not even an honest consonant to cover her tail feathers with. There are always others worse off.

16   That Fellow Acronym
         He all Time Make Trouble

Each of us has read sometime somewhere that in New Guinea pidgin the word for "piano" is (I use English spelling) "this fellow you hit teeth belong him he squeal all same pig". I am inclined to doubt whether this expression is authentic; it looks just like the sort of thing a visitor to the Islands would facetiously invent. But I accept "cut grass belong head belong me" for "haircut" as genuine.* I used to live in Shanghai at a time when Chinese pidgin was still commonly used there, and I can think of a few similarly amusing circumscriptions. For instance, the earthy word "to belch" was rendered as "inside come upstairs" and a tailor's model was referred to as "man no can speak".

Such phrases seem very funny to us, and make us feel very superior to the ignorant foreigners who use long-winded expressions for simple matters. And then it is our turn to name quite a simple thing, a small uncomplicated molecule consisting of nothing more than a measly 11 carbons, seven hydrogens, one nitrogen and six oxygens. We sharpen our pencils, consult our rule books and at last come up with 3-[(1,3-dihydro-1,3-dioxo-2H-isoindol-2-yl)oxy]-3-oxopropanoic acid. A name like that could drive any self-respecting Papuan to piano-playing.

It so happens that the pidgin languages have a lot in common with chemical nomenclature. Both deal with transient vocabularies (our mythical Islander is only briefly interested in the piano, else he would learn its name; and the chemist prepares his compound, finds it does not cure cancer, publishes its melting point and moves on to other things). And both systems of communication try to build up their transient vocabularies from a very limited stock of morphemes.

I beseech all my readers to make a mental note of this term "morpheme", for a great to-do will be made about it in the next Chapter. Morphemes are

---

* I have been informed by an expert that the actual phrase is *rousim algeder gras bilong hed bilong me*. *Rousim* no doubt comes from the German *raus,* "off with it!"

the smallest possible groupings of sounds or letters which convey information. Thus, the three-word title of this book contains six morphemes. *The* is the definite article; *Chem* means "the Science of Transformations"; *ist* means "one skilled in"; *'s* means "belonging to"; *Engl* means "the territory or culture of the Angles and Saxons"; and *ish* stands for "having to do with". Morphemes are thus to the linguist what atoms are to the chemist and, to complete the metaphor, we might say that the sounds ("phonemes") or letters are the protons and neutrons making up these atoms.

Now let us do some rapid and approximate calculations. In a fully developed language, the number of morphemes tends towards a limit of 10^5. The vocabulary available for common use (i. e. specialist nomenclatures such as those used in chemistry and biology do not count) is of the order of 10^6. I derive this number as follows: a very complete English dictionary is likely to have about 4×10^5 entries, but most of these can be modified by plural or adverbial endings, verb tenses etc., to give further words. Hence 10^6. A language, of course, goes on growing, but not uncontrollably so. For every three new words that enter the vernacular, perhaps two follow *fain* and *forsooth* into oblivion.

In chemical nomenclature the situation is quite different. The number of morphemes is of the order of 10^3 but as for the vocabulary — how many chemical compounds are there? By the time this appears in print, the number of compounds listed in *Chemical Abstracts* is likely to be somewhere between five and six million, but this is a vocabulary whose growth is almost uncontrollable. Unless the world is blown apart, the total number of chemical "words" is bound to reach the order of 10^7. It follows that the number of morphemes per chemical name is bound to be exceedingly high — a chemical name has every right to be a hundred times as long as a word in common use.

To drive this point home, let us take the ordinary word "table" and rename it according to chemical nomenclature principles. We would come up with something like "rectangle-r-1,c-2,c-3,c-4-tetrastick" (*r* stands for reference, *c* for *cis*) and if this had to be mentioned frequently in an article we were writing, we would be greatly tempted to abbreviate it to RATS.

RATS of this nature, unfortunately, infest our communications system in unbearable numbers. This type of abbreviation, called acronym, in which the initial letters of the parts of a compound noun are run together, is being abused beyond endurance by the headline writer and the snob who wants to create "in" jargon. Suppose a visiting American dignitary were to take an evening off from his official schedule to relax at King's Cross:

"USA VIP goes AWOL" would the headlines howl next day. Do we have to live with that kind of language, or can the RATS be chased back into the sewers?

We can at least try to keep the plague from spreading. Let us take common-use language first. Here, the golden rule is that if not all your audience is likely to be *instantly* familiar with the acronym, its use is an unforgivable discourtesy. Second, the abbreviation should really abbreviate. In my country, at the moment, the government is formed by the Australian Labor Party. This is sometimes abbreviated to "A. L. P." by broadcasters, and sometimes to "Labor". The latter is much to be preferred – it makes sense and has only two syllables to the acronym's three (ay-ell-pee). Then, the expression abbreviated should occur reasonably frequently; there is no point in creating an acronym for a compound noun that occurs twice in a long article. If these conditions are met, I am willing to live with a reasonable number of acronyms.

(One of these is FASEB, the Federation of American Societies for Experimental Biology. Its Director of Publications, Dr Karl F. Heumann, who shares my fascination with language and has made many valuable suggestions for this book, was the first to point out to me that American usage restricts the term *acronym* to those abbreviations that can be pronounced as a single word. I take note of this, but shall continue to use *"acronym"* in the sense of "any combination of first letters in a word sequence". Pronounceability is a relative thing, and a Czech might well get his tongue round a sequence of letters that would leave an Anglo-Saxon tongue-tied.)

In the language of chemistry, as our calculation has shown, the pressure to create acronyms is almost irresistible. The best an editor can do is to restrain and channel the flow. My colleagues and I of the *Australian Journal of Chemistry* have a profound aversion to acronyms consisting of full capitals, and do away with them whenever we can. We have three reasons. First, in a surprisingly high number of cases the acronym presents no great gain of space over the formula: DMSO is just as wide as Me_2SO, TMS has only a negligible advantage over $SiMe_4$. Second, the acronym could easily be mistaken for a formula (DMF might be a compound of metal with deuterium and fluorine). And finally and most important, a succession of full capitals in the middle of the text offends the reader's eye; we feel that their appearance RIPS THE PAGE APART.

Fortunately, in our fight against the full-capitals acronym we have a worthy ally in the Commission for Inorganic Nomenclature of the International Union for Pure and Applied Chemistry (I must admit I generally call this organization IUPAC). In Rule 7.35 it is clearly stated that organic ligands in the formulae of coordination complexes must be abbreviated by use of lower-case symbols only. We enforce this rule with fanaticism, and are most grateful to the Commission for doing their aesthetic duty. Their colleagues in biochemistry permitted the use of full capitals for acronyms representing enzymes, coenzymes and other metabolites, and the lamentable results are all too visible in every biochemical paper.

To help its readers to comprehend abbreviations quickly, the *Australian Journal of Chemistry* uses a system of "like symbols for like concepts". Thus, we print a succession of lower-case letters for acronyms of chemical compounds (except for such old standbys as acac, phen, en etc. these should be redefined at the beginning of each paper). For techniques, the lower-case letters of the acronym are separated by full points (n. m. r., e. s. r., g. l. c., t. l. c.). For computer programs and for the esoteric abbreviations which only molecular-orbital theorists understand, we use an alphabet which the printer calls "small caps". This consists of letters having the shape of full capitals but only the height of the standard lower-case letter (the printer calls this "x-height"). Their appearance in mid-sentence is not as aesthetically offensive as that of full capitals. Yet the careful editor will not allow them unrestricted access to the printed page; the reader will tolerate a limited number of these symbols but will get IRRITATED IF THEY OCCUR TOO FREQUENTLY.

Considering the great number of acronyms that are being proposed for public use, I find it surprising that so few of them gain permanent acceptance. Certainly, a good many technical terms, of which *radar* and *laser* are the most prominent, have entered the language in the second half of this century and will last as long as our present technology lasts. But non-technical words? Certainly we could not do without *W. C.*, and *TV* seems headed for permanence. (I note in passing, though, that Italian newspapers have taken to writing this as *tivù*, to make it seem less of an acronym; and come to think of it some sections of the English press deal out the same treatment to *emcee* and *dejay*.) But what else is there? Etymologists are by no means certain about the origin of the word *posh*. I have often heard it said that the word *cabal* was made up of the initials of five intriguing statesmen, but this is not so: it comes from the Hebrew. The word *jeep* is disappearing from the vocabulary, and the rather delightful wartime word *snafu* is seldom seen these days, and not for want of opportunity. Clearly acronyms face an uncertain future; do not lightly add to their number.

17 On the Divisibility of Earth/Worms

I published my first study of the earth/worm problem in September 1976. Now, as I sit down in May 1984 to revise this contribution, I reflect on the considerable changes that have taken place in less than eight years. In some respects (the burrowing of earth/worms into everyday language) the problem has become far less acute, in others (the influence of the earth/worm on the Chemist's English) it has become more grave. It seems to me that a mere updating of the original article will make this chapter look like patchwork. What I propose to do, instead, is to reproduce my original article, and then follow this up with a postscript explaining the changes. The very fact that, in a book of this nature, a chapter may need radical revision after a few years will interest students of the Chemist's English and deserves some comment. Here is, first of all, what I wrote in 1976.

17.1 The Original Article

Suppose that you have just finished your fundamental treatise *The Extraction of Ytterbium Oxides from the Sands of the Simpson Desert by Means of Specially Trained Earthworms*. The appropriate journal for a paper of this nature is *Biometallurgica Acta* which, as everyone knows, is published as a public service by the firm of Quickquid and Fastbuck (editorial and advertising offices in Vaduz, printery – for tax reasons – in Helsinki). Your paper is accepted and in due course the proofs arrive. You note that the word "perchlorate", at the end of a line, has been divided thus: "perch-lorate". You demand that this be corrected and later receive a letter from the Finnish printer: "Honoured Professor! Can you inform me of the principles according to which, when it occurs at the end of a line, an English word should be divided?"

All right, how are you going to answer? I can tell you in advance that it's no use looking up your high-school grammar; nor will you get any help from your English (as distinct from American) dictionary. The question of word division is the most scandalously neglected problem in English teaching, and yet it is of enormous importance. If the printer of this article were to use a word division which our alert production editor found unacceptable, we would be up for the expense of resetting *two* lines, rather than one. In fact, the vagaries of word division are the principal reason why "computer-setting" of newspapers has not yet become economically feasible, and various publishers are trying to overcome the difficulty by feeding an entire American dictionary into their computers — for American dictionaries do care about consistent word division.

This gives me a chance to renew my guerilla warfare against the publishers of English dictionaries. They have caused this trouble by their attitude that these books are meant for scholars (who with their goosequills are free to indite lines of unequal width on parchment) and not for such menials as typists and printers. Not till 1974, to the best of my knowledge, has the great Clarendon Press, over whose imprint all the Oxford Dictionaries appear, seen fit to publish an American-style lexicon (for "Advanced Students"). In the meantime, the Americans have dominated this linguistic field, and one way to answer our Finnish printer is to ask him to read the appropriate section of the Introduction to Webster's *Third New International Dictionary*.

The trouble with the suggestion is that this section is fiendishly difficult. In trying to fight my way through this maze of complex linguistic lore I have, after every paragraph, felt like turning to a little paper on quantum mechanics for light relief. Before I share such knowledge as I have gained with you, I should like to tell you the simple way in which we of the *Australian Journal of Chemistry* solve our problems of word division. In what follows I shall use the oblique stroke ("solidus" says the printer; "slash" says the computer programmer) to indicate word division, in order to distinguish my example from true hyphens and from word divisions made during the setting of this article.

In consultation with our highly intelligent and literate friends in the keyboard room of our printery, we have evolved the following system:

1. All "common-usage" words are to be broken as indicated in *Webster's Collegiate Dictionary* (note that we follow *Webster's* for word breaks but the *Concise Oxford* for spelling).

2. In words that are in the domain of the Chemist's English we overrule *Webster's* and divide at the morpheme boundary in *all* circumstances (i. e. we break the word according to etymology even if this disagrees with pronunciation). Our printers have a long list of such words; examples are *hydro/lysis, chromato/graphy, thermo/meter, sulf/oxide* and even *nitr/amine*.

3. For formal chemical nomenclature the printers have a memorandum which tells them were "breaks" are most likely to be found. Thus it is safe to break after *yl*, after *o* if it is not followed by *x*, after *y* if not followed by *l*. In the case of specially difficult words, we indicate the "best breaks" by gentle pencil strokes: *chloro/penta/fluoro/benzene.*

This system works very well and has cut down the number of corrections; I advise all my readers who have to prepare copy for publication to come to a similar understanding with their typists or printers. But no doubt you are not content with remaining helplessly at the mercy of a dictionary, and would like some insight into the basic rules of word division. For this, we must revive the term "morpheme" which in my previous article I defined as the smallest possible assembly of letters or sounds which convey information. Most other European languages respect morpheme boundaries only in the case of compound nouns which are joined together, but in English respect for this boundary goes much further, according to prevailing custom (note that in English *prevail/ing* is broken at the morpheme boundary, whereas a French printer would break *préva/lent* with pronunciation). The main rules are:

A) Where morphemes are joined together in such a way that the pronunciation of each component is not greatly changed, the morpheme boundary is *re/spect/ed.*

B) In other cases, pronunciation prevails. Thus *pre/vailing* becomes *prev/alent, re/fer* becomes *ref/erence.*

But of course "pronunciation prevails" is a whishy-washy term. I have a continuing battle with all typesetters who join our printery over the word *corre/spond,* which, until their first encounter with me, they all attempt to break between the *s* and *p*. Some words are differently pronounced in different regions (e. g. *cyclic,* which *Webster's* breaks after the *y,* and even *methyl*). I shall try to take the term "pronunciation prevails" a step further, but must warn everyone that what follows will be a very sketchy treatment of the subject. Let us use the symbol V for a long vowel (or a long diphthong such as *ow*), v for a short vowel or diphthong and c for a consonant ($c_1 c_1$ means a doubled consonant, and $c_1 c_2$ a pair of unequal ones). The prime sign ' indicates that the vowel which precedes it is stressed. The customary divisions then are:

$$\ldots V/cV \ldots, \quad V/cv, \quad V/c_1c_2v, \quad Vc_1/c_2c_3v$$

$v'c/v, \; v/cv'$ (if both short vowels are unstressed, then v/cv), $v/cV, \; vc_1/c_1v,$
$v'c_1/c_2v, \; v/c_1c_2v, \; vc_1/c_1c_2v, \; vc_1/c_2c_3v$

$$cv/Vc, \quad cV/vc$$

This seems to take care of most possibilities very neatly, but unfortunately things are not quite so simple. Remember, first of all, that all these rules only apply when the "morpheme boundary rule" (A), given above, is not invoked. And then special cases abound. For instance, groupings of letters which represent only one sound are counted as one letter (*sh,* or *ti* in na*ti*on; *ch,* which is not really one sound, is included in this group). Then, were *i* has a consonantal *y* sound, the word that contains it is divided just before this *i (Austral/ian),* contrary to the rules just given.

In general, though, those of my readers who need to know more about word divisions will find the above symbols a very present help in trouble. The main difficulty, really, is to know when to apply the morpheme rule A and when the pronunciation rule B.* My personal advice is: if in doubt, follow the morpheme rule A. If you have erred, at least you will have erred on the side of scholarship. And if you are dealing with Chemist's English, rather than everyday English, then throw the pronunciation rule to the winds and follow rule A. The Chemist's English is the language not only of the Universities of Sydney, Cambridge, Chicago, Auckland and Toronto, but also of those of Nigeria and Himachal Pradesh. Pronunciations may vary, but etymology remains. And thus let us have *thermo/meter* and not *thermom/eter,* Webster or no Webster.

This would make quite a stirring finale to my article, but for completeness' sake I must still explain what happens when a consonant is doubled in front of a suffix, as in *hit, hitting* or *stop, stopped.* In such cases, the break always occurs between the consonants *(hit/ting).* If a doubled consonant was already present before the suffix was joined, rule A prevails and the

---

* I am sure this will infuriate my readers, but I must mention that there are exceptions even to rule A. It applies to all compound words, and to all prefixes that leave the pronounciation intact, but in the case of suffixes there are exceptions. Thus */ing, /ed, /er* and */able* follow the rule, but *ous, ent, ence, ant, ance, or* do not. I don't know why.

doublet remains undivided *(miss/ing)*. If an *e* is elided before the suffix is joined, rule A operates *(ris/ing)* except where this would leave three consonants in front of the suffix. In this case the break is c_1/c_2c_3 or c_1/c_1c_2 *(spar/kling)*. You may be rather *puz/zled* by this, in fact it sets the mind a-*bog/gling,* but every language has its illogical little peculiarities, and English more than most.

17.2   And Now for the 1984 Postscript

I should really have had sense enough to foresee that what I wrote in 1976 would be overtaken by events. In the meantime ways have been found to "feed the dictionary into the computer" so that it recognizes the spots at which words in common usage can be divided. All major newspapers are set by computer typography now, and in all printeries the store of word divisions is continuously updated and perfected – whenever the computer comes up with an unsatisfactory word break, this can be suppressed by the operator and permanently replaced by the better version. Thus changes will be rare, and where they occur they will be inexpensive.

But this happy situation applies only to everyday or literary English. Thus I do not think that the revolution in typesetting has made the contents of my original article a luxury rather than a necessity. First of all, you may be preparing an article for a Journal that does not typeset but demands camera-ready copy, or you may be working on a typed report. And then, the present system very effectively protects you against that rapacious subspecies *eart/hworm* and (since this is likely to be in the computer's memory) *perch/loric,* but you are the only one who can protect *chloro/penta/ fluoro/benzene* from injury. And if your article is going to be typeset by a commercial printer who publishes scientific material only as a side-line, you may have to battle to make your preference of *thermo/meter* over *thermom/eter* prevail.

The advent of computer typesetting has been a most marvelous thing to witness. The coarse and crude approximations of earlier days have given way to superb systems capable of creating the most aesthetically pleasing line and the most helpfully laid-out page. But there is the inevitable price to be paid for technological progress; honest craftsmanship is displaced. The computer makes things too easy for the keyboard operator, and thus downgrades his work. I doubt whether future editors and authors will receive the support I have enjoyed from well informed printery foremen – people who have taught me more about the Chemist's English than I can ever properly acknowledge. Those of my readers whose publishing career is just beginning should know that they can expect only limited help from keyboard operators. It is all the more important, then, that they should be vigilant in the protection of our technical language, and intolerant of any improper segmentation of the earthworm.

18   Instant Stylistics

"Please tell us", say all my correspondents, "how to write a good paper!"

Now it is one thing to pontificate on participles or discourse on diagrams, but can "good" writing be taught? What is a "good" paper anyway? And why should an editor be an expert on the "good" paper? His traffic is with the "acceptable" paper, his job the prevention of mistakes. But mere absence of mistakes does not make a paper "good", surely?

18.1   How Good is "Good"?

For all it is worth, I shall offer my own definition of the "good" paper. It is a paper written so as to *yield up its information to the reader in the shortest possible reading time.* Let us postulate that you have stored in your notebooks data and conclusions containing a definite amount N of significant new knowledge. Our task will be to impart this knowledge to an interested reader of appropriate background in the reading time t, which should obey the condition that N/t is a maximum.

(Observe that our problem is not that commonly posed in information theory: what is the maximum speed with which a message can be *transmitted?* We are not concerned here with the time it takes to place the mental food before the reader, but with the time he needs to digest it; not with information transmitted, but with knowledge acquired.)

Now let us transform our idealized reader, by a cloning process, into a thousand identical readers and set them perusing a thousand different versions of your paper (all containing the knowledge N), all printed in the same type size. We note the time t needed by each clone to acquire the knowledge N, and then plot N/t against P, the area of paper on which each version is printed (Figure 1).

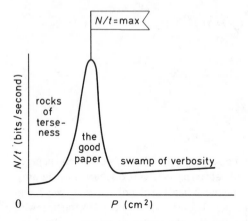

Fig. 1. Characteristics of written communication.
N = knowledge; t = time needed to acquire knowledge;
P = area of paper needed to transmit knowledge.
The origin of the graph represents communication by telepathy − it saves paper, but takes an infinitely long time!

The peak N/t = max will be seen to arise majestically out of dreary flatlands. To its right stretch the dismal marshes of verbosity; the reader struggles forward but is held back at every step by clinging verbiage and a porridgy surfeit of data. At the peak's left lies the rocky plain of excessive terseness; between the cruelly serrated rocks lie patches of quicksand, in the form of sentences in which the reader submerges without hope of extricating himself. Not surprisingly at all, our landscape spells out the message that excessive terseness and excessive verbiage alike are harmful to the reader.

Terseness and verbosity are matters of such importance that I shall dedicate separate chapters to them. Here, let us return to our quantities N and t, and follow our reader (shrunk back by retro-cloning into a single exemplar) in his progress through the paper. The knowledge N is now a variable; it increases in some way as t increases. If N is to be plotted against t, can we take it for granted that the graph should follow a straight line, $N = kt$? And if so, what should be the value of the slope k (which obviously represents the reader's capacity of comprehension)?

In the case of our idealized reader (I find I cannot relate to such a vapid abstraction − let's call him Joe) there is no question but that, *ideally,* the plot of N against t should be a straight line. Joe is unwaveringly attentive toward your paper and persistent in his desire to acquire the knowledge N; offering knowledge to him is very much like propelling a train on a flat track of constant friction. Obviously a constant speed is the most advantageous way to make progress.

18.2   Getting through to Joe

But can such a constant speed be achieved? The answer is a thundering "No", and here this article switches from generalities to practical issues. Some of the sentences in your paper will cause Joe's N to leap upward, others (such as the brand of your n.m.r. instrument and the method of purifying your solvents) will not increase N significantly but will only consolidate N already acquired.

You cannot leave this information out, and you also owe Joe chapter-and-verse proof of your statement (accompanied by an upward bound of N) that examplamine reacts with paradigmol in an atmosphere of xenon but not of argon. For this proof, you have to refer Joe to Table 4, which is a ponderous accumulation of numerical data garnered from 73 experiments. Joe will dutifully turn to Table 4, and spend 68 seconds examining it; yet, his N will increase only marginally over the value it acquired in the 1.5 seconds it took him to read your original, sensational sentence. And then there is the question of the by-products you obtained during your first, unsuccessful attempt to synthesize exemplaric acid. Their crystals glisten, their analyses are correct, they clamor to be included: yet these data advance Joe not a whit in his quest for the knowledge N he is after.

From this emerges some practical advice. Digressions, consolidating evidence, and similar matter should be kept out of the mainstream of the paper. They should be banished to those sections of the paper that are not intended for continuous reading – experimental sections, tables, and appendices. In every description of synthetic work, there will always be a few compounds that have not been invited to the party, but are gatecrashing in the hope of being introduced to Mr Beilstein or Mr Gmelin. Reviewers are generally tolerant, but make sure you don't let too many such compounds come along.

There are, of course, digressions that cannot be kept out of the main text. Without a brief description of your first unsuccessful synthesis, the second successful one could not be understood. There the golden rule is that *a digression should be firmly labeled a digression.* This can generally be done by skillful titling of subsections, or sometimes by just telling the reader where the interruption begins and ends. Joe is an unfailingly courteous man, but even he can get annoyed when he finds he has been lured off the main path.

18.3   How Steep our Slope?

Back to the mainstream: We have established that digression should be tidily served up as a side dish and that in the main part of the paper we should maintain the plot of N against t on a straight line of slope k. What is the ideal value of k?

To answer that question I must first recall the definition of the unit of information. A "bit" is the amount of information that allows you to decide between two equally likely possibilities. If a question is so framed that it can be answered yes or no, then by answering it you offer the questioner one bit of information.

Now it is my belief that the value of k should be near one bit for each main part of a sentence (a clause of substantial length, which Joe reads in about one second). Specimenane does not react with McInstance's reagent (yes/NO); therefore, its chlorine is covalently bound to its cobalt (YES/no).

Information scientists will violently disagree with me, and want to assign 20 bits to McInstance's reagent alone. But I shall firmly reply that I am talking not of information transmitted but of knowledge acquired; in such matters information scientists are less competent than psychologists.

Central to the above theory is the interested reader Joe, who is eager to read your paper. But in real life you are more likely to encounter his cousin Jim, who has a low boredom threshold and is just looking for an excuse to put your paper down unread. How can you hold this fellow's attention?

As an editor, I am not competent to answer; but I also used to be a playwright, and in the theater we had a saying; "In the first 10 minutes, somebody must kick the cat, or else the play will fall flat."

Well, go thou and do likewise. Kick the cat as hard as you can, somewhere in your introduction; "Classical theory affirms that modellane has aromatic character. Data we present in this paper show that this is not so."

That sharp "kick" in the graph (Figure 2) will gain Jim's attention and keep him going straight through your paper, or at least until the investigator from the Society for the Prevention of Cruelty to Animals arrives.

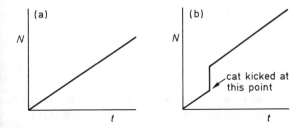

Fig. 2. How we acquire knowledge: Joe (a) likes to do it gradually, but Jim (b) must be kept awake by feline squeals (N = knowledge; t = time).

19 A Piece of Classified Information

I shall now try to make good my boast (see the preceding chapter) that I can assign a rough numerical value to the amount of information that Joe, the idealized reader of a scientific paper, can digest in a second. To do so, I shall have to rehash some elementary information theory and (do you remember my remark in Chapter 18 about kicking the cat?) give you a crash course in Chinese grammar.

19.1 How many Bits to a Message?

The information theory first. The "bit" is the amount of information (a yes-or-no answer) which decides between two equally likely possibilities. For four equally likely possibilities *A, B, C, D* you need two bits: you first ask whether the letter you seek is in the top half of the list, and ask your second question when you have received your yes/no answer. For eight possibilities you need three bits, and so on. The number of bits is the logarithm to the base of 2 of the possibilities, or the number of *b*inary dig*its* needed to label them (00, 01, 10, 11; 4 possibilities, 2 bits). There are 26 letters (i. e. 26 possibilities) in the alphabet of a telex machine, hence each letter would at first glance seem to carry just under 5 bits of information ($\log_2 26 = 4.64$).

Actually it is much less, according to the redundancy theory elaborated by Claude Shannon. As the message proceeds, the number of possibilities open to the next letter (or "signal") diminishes. After *str* there must be a vowel; after *q* there must be a *u*. In this article, after the letters *info* your mind would switch off until the start of the next word; the letters *rmation* offer you zero bits of info. (They are, as it were, on standby duty in case the first four letters become garbled.) A sensible estimate of the number of bits a single letter in an English message carries would be $1\frac{1}{2}$ bits.

All the above is of course a gross oversimplification; if the above sketch was all there was to the theory, we would be led to the conclusion that a message in an unknown language "carried" more information than one in English. What concerns us here is that Shannon, Wiener and other brilliant mathematician-philosophers have subtly succeeded in quantifying the amount of *information transmitted,* but numerical estimates of *knowledge acquired* are quite another matter. Psychologists conduct very meticulous "retention tests" (and modern psychologists express the results in "bits") but retention from scientific texts is something else again. How is a psychologist to assess the wisdom you have gained from reading a paper on molecular orbital theory?

19.2 Now for a Fast Trip to China

I thought it might be possible to make progress by using a word-by-word rather than letter-by-letter approach, and that made me think of Chinese. It is often said that Chinese has no grammar (actually it has a subtle syntax) but it certainly lacks the inflectional grammar of European languages. Verbs do not change to indicate the future, nouns have no plurals. All is made clear from the context. The Englishman says "He go*es*" or "I *shall* go tomorrow". The Chinese considers the italicized parts redundant.

Chinese, however, has a grammatical feature which we lack. This is group of words called "classifiers" which are inserted in front of the noun. There are about 40 of them, and they all would have to be translated as "piece"; but the first classifier can only be applied to persons, the second to animals, the third to stick-like things, the fourth to sheet-like ones, and so on. One piece$_1$ woman, two piece$_2$ cats, this piece$_3$ pencil writes on this piece$_4$ paper. The importance of the classifier in a language with the phonetic characteristics of Chinese is obvious; there might be dialects where there is little difference between the sounds of *pencil* and *paper*.

Now let us harness, for our own purposes, the ability of classifiers to group the words that follow into sets. Let us imagine that each English word is preceded by an invisible classifier. We shall now investigate the sentence:

Copper(II) salts are blue.

According to the letter-by-letter approach this might be worth about 40 bits, but of course if you address this statement to a stranger in the street he will acquire no knowledge thereby; at the best he might memorize it long enough to repeat it to the hastily summoned attendant from the mental hospital.

19.3   ... Back to Joe

Now let us imagine Joe, who is suddenly only 16 years old, seated at his high-school desk waiting for the inorganic chemistry lesson to begin. The teacher arrives and writes our sentence on the board. Joe knows that the main topic is inorganic chemistry, hence the number of possible imaginary classifiers preceding the first noun is limited, say about 10 ($piece_1$ for atomic entities, $piece_2$ for reaction groups, $piece_3$ for thermodynamic concepts, etc.).

Within the set "atomic entities" there are about 100 elements having (we are in high school, remember!) two oxidation states on the average. Say 200 members for this set. The word "salts" in our sentence could be preceded by at most five classifiers (acids, bases, hydrates . . .) and each set would have only one member. The verb "are" is so low on information content that the Chinese, Russians and a good many others scornfully omit it. "Blue" might have been preceded by about 10 classifiers (chemical properties, physiological properties, economic importance . . .) and in the set "color" there might be 20 members. Hence our sentence, as far as Joe is concerned, is one of

$$10 \times 200 \times 5 \times 1 \times 10 \times 20$$

or 2 million possibilities, worth 21 bits.

Can Joe digest this amount of information in the half-second it takes him to read it? Not a hope. Just imagine what confusion would result in his mind if the teacher were to go on writing, at a steady speed, "Iron(II) salts are green" and so on right through the periodic table.

19.4   The Soft Approach

How then does the teacher make Joe digest the knowledge contained in our sentence? He will not brusquely write it on the board but first address the class as follows: "Today, dear children, we shall talk about the properties of some compounds of metals in Group 1B." Check, and you will find that after such a preamble our sentence is one of about $2 \times 6 \times 5 \times 1 \times 10 \times 20$ or 12,000 possibilities; we are down to $13\frac{1}{2}$ bits.

But this is still far too much. To give Joe a chance, the teacher now adds: "You may remember, after some plants have been sprayed, that they look blue. This is because they have been sprayed with copper sulfate solution."

This might just do the trick. If the teacher is enthusiastic, he may add that it was one of the attractive achievements of modern Australian chemical research to discover (here there will be a stir of interest in the class) that the blue sheen on the leaves of some local plants arises from the presence in the leaf wax of substances which can combine with copper. At this, something will click in Joe's mind. The teacher's battle is won.

But the teacher has now been speaking for half a minute. To be sure, he has ladled out dollops of subsidiary information about plant sprays and leaf waxes. But these were just means to an end: to make Joe acquire the specific knowledge which at the outset we assessed at 21 bits.

19.5   1 Bit s^{-1}

Here you have my proof (of course it isn't a proof; it is just the vaguest of tentative estimates arising out of my very sketchy knowledge of information theory and dressed up to look scientific with a lot of prattle about China) that the reader of a scientific paper likes to ingest his knowledge at a rate not too far remote from one bit a second. I stress again that I am talking about *essential* information: subsidiary information and consolidating information will of course have to be supplied. The knowledge we discuss here is the one that *remains* with the reader after he has put your paper away.

Has all this any practical value? Perhaps. It may pay you to pick out the odd paragraph in your first draft and say to yourself: "Now this takes a minute to read. Does it contribute enough towards the final, essential knowledge that I have to offer the reader?" It may not contribute enough, because it is full of rambling asides. Or you may urge the reader up too steep a slope, say by introducing the latest fad term, still not widely known, from Woodward–Hoffman theory. In the previous article, a "good" paper was arbitrarily defined as one which could be understood in the minimum time. We can now say, more precisely, that a "good" paper is one in which the flow of information is smooth. Some would try to put it in a nutshell by saying a good paper should have *structure,* others would use the fashionable term *thrust.* My favorite saying is: it should have *drive.*

I hope this is of some help. But if, inspired by this article, you restructure your paper, and still have trouble with the referee over its presentation, don't blame me. Criticize Confucius.

20 An Investigative Examination of Driveliferous Jargonogenesis

The subject matter of the *Journal of Chemical Information and Computer Sciences* is efficient communication. Thus it is rather embarrassing that I should find myself choosing a paragraph from that particular journal as the initial target in my sermon against verbosity. It stands at the beginning of an otherwise excellent paper entitled *Chemical Abstracts Service approach to management of large data bases* and reads as follows:

> In the literature, largely inspired by the concepts in the SHARE/GUIDE documents[6,7,14] on data-base management and attention generated by the CODASYL Data Base Task Group[5], there has been much discussion of the function and organization of data-base management activities within an enterprise. Most of these publications, however, envision a model for information support to a business where the thrust of the business is something other than information – e. g., automobiles, general retailing, or some other commodity or service. These publications are good sources of techniques and principles, but they tell little about the development and evolution necessary to transform very large information handling operations into a "data-base orientation".

At the end of the paragraph the reader has a feeling of unease, of language badly used; how, for instance, can an "operation" be "transformed" into an "orientation"? One has to work one's way deeper into the paper before being certain of what the first paragraph meant to convey. It is simply this:

"Earlier papers[5,6,7,14] on the management of large data bases described the case of an enterprise accumulating information for its own use, to help it manufacture merchandise or offer a service. We shall now discuss the situation where the information itself is the end product to be offered to others."

Now the interesting thing about the lengthy paragraph I have quoted is that the authors can do much better when they want to; they are, in fact, among the world's leading specialists in the art of compressing information. After another overblown paragraph or two, the paper settles down to tell of the fascinating techniques by which *Chemical Abstracts* so slim down their data that they are able to keep a vast amount of them in the direct-access section of their computers. *Chemical Abstracts* people, moreover, are generally very good, brisk writers – over the years I have engaged in a good deal of correspondence with them and their letters have always been courteous, helpful, and crisply written. Then why this lengthy, nebulous opening paragraph?

The answer is, surely, that there are occasions when we feel that jargon is demanded of us. The problem of jargon is a social one rather than one of syntax. Human society divides spontaneously into groups, and each group is certain that it is the Rightful Aristocracy, the Band of True Believers. When the day of reckoning comes, the followers of magnetoballistics will stand at the right hand of the President of the Academy, but there will be much wailing among those who put their faith in piezographics. And so we generate jargon, to cement the bond between the true believers and to keep the infidels out.

Similar to jargon, the use of "in" words, is verbosity, the use of "fat" words. The longer the word is, it seems to us, the more social status it confers on its user. In English-speaking countries, above all, there exists a suspicion that anything that can be described by a short word must needs be coarse and primitive. It is not pure coincidence that the words which caused such anguish to our fathers all had four letters. Aunt Agatha, blushing, told Aunt Emily: "It is to be feared that an act of intimacy may have taken place". We use equally simpering language when we say: "The crystals were subjected to visual inspection". The shade of Aunt Agatha hangs over us when we cross out "study" and substitute "investigation", or when we substitute "in consideration of the fact that" for "because". Our writing thus becomes an apotheosis of pusillanimous polysyllabicity, and editors incarnadine multitudinous manuscripts to get rid of it again.

They have to, if they want to balance their budgets. Incidentally, did you recognize that word "incarnadine"? Macbeth, having murdered Duncan, groans that his hand will "the multitudinous seas incarnadine,/making the green one red". It is a line famous for its sonority; nevertheless it is a bad line. The over-erudite words do not fit Macbeth as well as they would have fitted Hamlet; and Shakespeare was well aware that he had talked over his audience's head, for he repeated the thought immediately in simpler language. Shakespeare got away with it, but don't be too sure that you will, if you use similar verbal extravagance.

How can you protect yourself against excessive verbosity, then? I suggest that you look at your first draft, to see whether you have committed any of these four grievous sins:

1. *Loudly saying nothing at all.* The citation of the acronyms in the first sentence of the paragraph I have criticized is an instance in point. It soon becomes obvious that they are not relevant; at best, they might have been mentioned among the references.

Let me give other examples, ranging from the oldfashioned

> One morning, returning to my laboratory at an early hour, I noticed that a solution of gurglionium porphyrate had, under the cold night conditions, shed a beautiful crop of azure crystals, many of them large and perfectly formed . . .

to its modern counterpart

> Careful examination by visual and spectroscopic methods revealed a certain number of characteristic differences between the electronic spectra to facilitate differentiation between the blue and green isomeric forms. It has been reported by several groups of independent workers that *cis* isomers have stronger bands with higher extinction coefficients than have those of the *trans* forms corresponding to them and it has been observed further that this difference is more prominent . . .

Any comment on this would be a further example of verbosity.

2. *Proudly proclaiming the obvious.* In everyday speech, occasionally, we "waffle" on; to regain composure, perhaps, when we have been disturbed and distracted. We come out with statements like: "There's never a villain dwelling in all Denmark/but he's an arrant knave". But there is no excuse for such bits of noninformation in writing. Some time ago, a correspondent took me to task for not writing about verbosity. He convinced me that the problem was serious by including, in one of nine hefty paragraphs, the sentence: "Nevertheless, it must be remembered that far from being a static language, English is a living language, which is continually evolving". Very true, but there needs no ghost, my lord, to tell us this.

3. *Saying the same thing several times over.* Look again at the first sentence of the paragraph I quoted at the beginning. One piece of information ("there have been earlier papers on the subject") is given three times: (a) by the words "In the literature"; (b) by the superscript numerals denoting references; (c) by "there has been discussion". Cutting "In the literature" would not have damaged the sentence at all.

4. *Pompous polysyllabicity.* Now please don't get me wrong; I am not against long words or phrases, but they must be used in their proper place. The short everyday words are like the common reagents you keep on your shelf, the esoteric long words like the expensive substances you lock in your cupboard. You would not think much of a chemist, would you, who used osmium tetroxide for a reaction in which sodium hydroxide would perform just as well? And yet the right occasion may arise when the precious reagent is brought in triumph from its hiding place, to resolve the difficulty it alone can resolve. And then everyone acclaims the master synthesist. Or (if we substitute words for reagents) the master stylist.

Have you found the marks of the four grievous sins in the draft of your paper? If so, erase irrelevance, omit the obvious, take out tautologies, and pare off pomposities. And remember that sentences built around active verbs are shorter.

I began this chapter by quoting a paragraph of verbose writing; I shall close it by quoting scientific writing at its best. The paragraph that follows was pointed out to me by Professor D. P. Craig, for whose help in preparing this article I am most grateful. It is from *Quantum Mechanics* by Dirac:

> When we make the photon meet a tourmaline crystal, we are subjecting it to an observation. We are observing whether it is polarized parallel or perpendicular to the optic axis. The effect of making this observation is to force the photon entirely into the state of parallel or entirely into the state of perpendicular polarization. It has to make a sudden jump from being partly in each of these two states to being entirely in one or other of them. Which of the two states it will jump into cannot be predicted, but is governed only by probability laws. If it jumps into the perpendicular state it passes through the crystal and appears on the other side preserving this state of polarization.

Crystalline prose, through which the meaning radiates unabsorbed and undiffused! If Dirac can be so lucid on quantum mechanics, do you and I have to muddle our account of the diazotization of examplamine?

21 *Brevity = Soul of Wit?*

In Ashburton, a south-eastern suburb of Melbourne, lies Y Street, so called because of its shape. Its law-abiding inhabitants (they had better be law-abiding; 1 Y Street is a police station) have every right to congratulate themselves on their choice of domicile. Apart from enjoying the protective proximity of the Law they are also only a minute away from the railway station, and within easy walking distance of shops, schools and parks. But these are common enough benefits; what really sets the denizens of Y Street apart from the rest of the metropolitan population is their ability to fill in forms rapidly in quadruplicate. When it comes to purchasing a freezer on easy terms, the man from Y Street is likely to have his application for credit approved before the one from Old Diamond Creek Road is even half-way through his first form.

Does this induce you to go cruising through the street in search of vacant houses? Think again. Perhaps you receive a great deal of mail; in that case, can you be sure it will reach you? A street name which consists of a single letter can easily be made unreadable by a stamp or blot; moreover, there are people whose handwritten Y resembles the numeral 7, so that 23 Y Street might be misread for 237 Street. To think that important communications might languish in the Dead Letter Office while you await them impatiently!

And here we are back at the Chemist's English because I feel I should balance my sermon against verbosity in the previous article with one against exaggerated terseness. In Chapter 18 I defined the "good" paper as the one which yields up its information to the reader in the shortest possible time. This led to a denunciation of fatty and pompous prose, of irrelevancies and of duplicated statements. All these slow the reader down. But if, zealously, you trim the fat off your first draft, your knife may slip and you may remove some muscle also. In that case, too, you cause the reader to lose time, for he will have to read your sentence several times over before he sees the connection.

And he may be just as annoyed with you as if your sentence had been overblown. Overterseness is often a symptom of snobbery; what business, the author seems to be scornfully asking the reader, have you to read my immortal treatise if you don't even know that SEDDG stands for self-effacing downward-decoupling gyromechanism? There is about the overterse writer the rudeness of a man who charges straight down the street without the slightest regard for his limping companion.

Now you don't want to be snobbish or rude. To protect yourself against being thought so, always ask yourself the question of whether you are going too fast or too slow for your companion, the reader. In particular, make sure you have not committed, involuntarily, any of the following discourtesies:

1. *Wrongly assuming ideas to be self-evident.* – This is particularly apt to happen if your paper is in a borderline field, say somewhere between organic chemistry and biochemistry, or inorganic and physical chemistry. You may write, in all innocence:

> These results show that qualunquine is eliminated by *Exemplar paradigmaticum* in the form of dioxoqualunquic acid. This disproves the theories of Nimportequi on glupid cell metabolism.

This may be immediately meaningful to a biochemist. But you are publishing in an organic-chemistry journal because your paper deals with the identification of the degradation products. Your readers may not know that the genus *Exemplar* belongs to the family Glupidae, and they may not have heard of Nimportequi and his belief that the plant cell would turn qualunquine into its hexahydro derivative. It may be that your last sentence is not relevant to your paper and should be eliminated. But perhaps it is the very sentence that holds the paper together; in that case make sure the reader does not stare at it in bewilderment.

To give another example: We all have, at some time or other, seen the words "it follows that" on a separate line between two complex mathematical expressions, and have been unable to see why and how "it" follows. The transformation linking the two expressions was not as self-evident as the author thought, and he should have given the reader a hint, perhaps in the form of an appendix.

2. *Name dropping.* – There is, of course, no need to "explain" such names as Diels–Alder and Woodward–Hoffman. But if you use, say, the Hunsdiecker reaction, a quick word of explanation or a reference might not be amiss.

3. *Acronym dropping.* – This can be particularly infuriating, because snobbery is immediately suspected. No editor will expect you or allow you

to define such common standbys as n.m.r. and t.l.c. But when it comes to the very latest decoupling technique or mass spectrometric process, be careful.

4. *Obscure jargon.* – A similar situation again. Such words as *antarafacial* and *synperiplanar* are now part of the Chemist's English and need no defining. But other words have not caught on so well. The correct thing to do is to redefine (at least by giving a reference) a new term of doubtful currency. Better yet, steer your way around the obscure word. It may not be as indispensable to the language as you think.

5. *Hieroglyphics.* – It is a very ugly, and unfortunately very common, practice, to make the main verb in a sentence a mathematical symbol, e. g.

$$K > 0 \quad \text{only if} \quad T > 350 \text{ K} \tag{1}$$

You may say this is a succinct and effective way of rendering the thought: "The factor K is positive only if the temperature is greater than 350 K." But has this thought been unambiguously expressed? Let us write our sentence again, with a small initial addition:

$$\text{It was found that} \quad K > 0 \quad \text{only if} \quad T > 350 \text{ K} \tag{2}$$

Now the symbols have not changed at all but suddenly they signify the past tense: *was positive* and *was greater*. Sentence (1) states an unalterable truth, valid for no matter what range of temperatures. Sentence (2) sounds far less certain: the authors have established a limited truth for a limited experimental range, but for all they know K might reverse its sign yet again if the mixture was supercooled.

What I am driving at is this: a sentence is a delicate piece of machinery, and the main verb is its most sensitive part. Now it is simply barbarous to remove this finely tooled piece of equipment and replace it with a crude symbol which cannot differentiate between singular or plural, between present or past, between the certainty of an indicative and the doubt of a subjunctive – and to call the result a sentence. No rage $>$ that of a sensitive reader who is exposed to this kind of non-communication.

Let me qualify what I have just said: my objection is only to the replacement of the *main* verb with a symbol; in parenthetic or peripheral clauses this practice may do no harm. Also, I mean no offence to the language of mathematics. Everybody agrees that mathematics is a language and, just like English, it has a grace and beauty of its own. But when two graceful and beautiful languages are jumbled together, the result is a graceless pidgin. I shall quote an example, carefully disguised, from a manuscript we actually received:

> From equation (17), when $n = 2$, $r = 3$ and the expressions (6) and (8) yield values similar to those previously reported.

How long did it take you to discover that the key verb of this sentence is the second = sign? Incidentally, when a colleague of mine tried to amend this sentence, the author was most indignant.

The example above reminds me that numerals denoting structural formulae may also turn into hieroglyphics in the hands of a careless writer. Do not write, for instance:

> (9) reacts with (13) to give (14) in 17% yield

especially if it is doubtful whether the structures will appear on the same page. In courtesy, you ought to refresh the reader's memory by writing: the acid (9) reacts with the amine (13) to give the amide (14) . . . Your sentence will not only become comprehensible, it will also look better. Just as mathematical symbols are "miscast" as verbs, so numerals "don't look the part" as nouns.

At the end of this chapter, the equation that forms its title remains unconfirmed. Brevity, if it is stylish, will always be the soul of chemical wit, but not if it is rude and snobbish. Let me replace our equation by another relation, for the accuracy of which I can vouch after much testing in the editor's laboratory:

> brevity $-$ wit $\rightarrow 0$

In other words, brevity of the witless type achieves nothing. Except to create more entropy; and we have got enough of that already.

22 *One, Hand, Clapping*

Imagine that you are editor of a scientific journal. The manuscript that you are just marking up for the printer contains several mathematical expressions. Your pen moves routinely through these, but suddenly it comes to a stop. You have just noticed the following equation:

$$3a - b) + c = 0$$

Clearly this it not right; and yours is now a common editorial predicament. The expression in front of you is meaningless, and it is up to you to find out what the author's meaning was. There are two possibilities. It may be that the unpaired) sign − "closing par", the printer calls it, "par" being an abbreviation for "parenthesis" − is a copying error and should not be there at all. Or else, naturally, the "opening par" (is missing.

Now editors have a sixth sense for distinguishing between authors' and typists' mistakes. You look for all the telltale signs and become convinced the typist is not to blame: the "closing par" sign has been introduced deliberately. The next task is to find where the missing (should go. This causes little trouble; the only logical place for it is between 3 and *a*. An opening par preceding 3 would make no sense, for the pair of parentheses, in the context, would then perform no operation and be just "noise". The same reasoning rules out $3a - (b)$; and $3a(- b)$ would be an ambiguous notation not in harmony with the rest of the paper. So, grumbling over the three minutes or so that you have lost because of the author's carelessness, you write $3(a - b)$ and prepare to go on to the next line. And at that moment, doubt strikes. The rest of the manuscript is in excellent order; how could such a gross error have gone unnoticed? Has someone, perhaps, invented a symbolic notation that you have not heard about? Sighing, you reach for your telephone: "Sorry to bother you, but would you confirm a small correction? You seem to have slipped up in your equation 24 . . ."

"Haven't slipped up at all" comes a self-assured voice over the phone. "Any fool knows that a closing par presupposes an opening par, and if there is only one obvious place where it could go, why bother to put it in? If *you* could figure out the meaning of the equation, do you think my readers, who are specialists in the field, will get stuck?"

You scratch your head. You had heard tales that this man was getting a bit eccentric, but such arrant lunacy! You mumble something like yes, ahem, you can see the point, but a journal must have a house style, mustn't it, and our house style is that all parenthesis signs must come in pairs. "And frankly, old man" you conclude "a par sign without its mate, I think, is just as useless as one hand clapping!"

I end my scenario here, fading out on the sight of you muttering angrily to yourself. But now comes a dramatic surprise. The sin against good communication I have just described, which would infuriate you no end if you saw it perpetrated in the language of mathematics, is one that you (and I and all your colleagues) often commit when writing English. Take the sentence:

> It is obvious that under the forcing conditions used by Wallach
> in 1896, formation of an intermediate could not have been ob-
> served.   (1)

Here the solitary comma performs very much the same function as the unpaired) sign in the equation: it indicates the end of a parenthetic expression. For reasons of logic and symmetry, the "closing comma" that ends such an expression should be preceded, at the point where the expression begins, by an "opening comma". Thus we should write:

> It is obvious that, under ... 1896, formation ...   (2)

When I insert an "opening comma" into a manuscript the author is often surprised and occasionally he asks whether the comma should not have been placed in front of, rather than after, the word "that". Most writers insert commas at the point where, had the sentence been spoken aloud, the speaker would have paused. But they only note the obvious pause of the "closing comma" and miss the shorter pause of the "opening comma". If you are writing in conversational English, as you would in a personal letter, this is quite forgivable. But in scientific texts one should be more careful. I shall give my own rule of good usage, and work appropriate examples into its enunciation:

"A parenthetical phrase occurring in the middle of a sentence need not *if it is simple* be separated by punctuation at all. If it is separated, as in this example, by commas or (as may happen in the case of lengthier phrases) by brackets then these symbols — and this applies also to dashes — should always *occur in pairs.*"

Observe that I am talking about parenthetic expressions in the middle of a sentence. If they occur at the beginning or end, one of the commas is lost. For instance:

Under forcing conditions, dehydration occurs.

So far, I have only discussed "parenthetic" commas. There are other types of comma that give chemists trouble:

Adjectival clauses. — Consider the pair of sentences:

The compounds which gave a positive Crump test were treated
... (3)

The compounds, which gave a positive Crump test, were treated
... (4)

The second sentence means that all the compounds considered reacted positively; the first means that only a part did. Purists recommend that in adjectival clauses of the first type, called "defining" clauses because they specify a subset, *that* be used instead of *which*. It is difficult to observe so subtle a rule in speech but, if you observe it in writing, it will enhance the quality of your prose. More of this in Chapter 28.

Commas preceding conjunctions (such as *and, but, if*). — These are not needed if the two clauses of the sentence separated by the conjunction are simple and of equal emphasis:

The isomers were formed in high yield and proved easy to separate. (5)

But, by inserting a comma, you can give a subtle emphasis to the second part of the sentence:

The isomers were formed in high yield, but were difficult to separate. (6)

A comma is mandatory if the first clause ends in a noun similar in meaning to that with which the second clause begins:

The liquid was introduced into the flask, and the dropping funnel filled with tetrahydrofuran. (7)

Commas in enumerations. – Whether the second comma is needed in *"A, B,* and *C"* has been much debated. Until a few years ago, the *Australian Journal of Chemistry* used to insist on this comma, but we then switched to *"A, B* and *C"* on the grounds that, where good usage permitted two alternative styles, the simpler one ought to prevail. But we apply our new "house rule" with common sense. Thus, if one of the elements *A, B* or *C* is further subdivided, the comma is reinstated (A, B_1 and B_2, and *C*):

> Reduction was attempted with Raney nickel, zinc and hydrochloric acid, and lithium aluminium hydride.   (8)

We also cheerfully break our own rule and insert the comma if the element *B* is lengthy. For instance, the comma would remain if in the above example $Zn + HCl$ were to be replaced by "the Jemand reagent prepared according to the method of Qualcuno".

Anti-garden-path commas. – There is one overriding rule which says: A comma is always welcome where it avoids confusion. Look again at sentence (7). Had its comma been omitted, the reader would have been propelled up the garden path for the time needed to read the next four words. I shall borrow two further examples from R. S. Cahn's admirable *"Handbook for Chemical Society Authors":*

> Unfortunately diazotized anthranilic acid decomposed . . .   (9)

> The ester dissolved in benzene was added to . . .   (10)

In example (10) the absence of a comma pair misleads the reader into thinking that *dissolved* is an active verb rather than a participle.

When I first became an editor, we had a rigid rule that *however,* if it occurred in the middle of a sentence, should be surrounded by a comma pair. I have long ago retreated from such rigidity, and nowadays just follow the author's preference. But the situation is different if the word stands at the beginning of a sentence. Take the expression:

> However examplitol was treated . . .

This sentence might conclude ". . . the tertiary hydroxyl remained" and *however* would then have the meaning of *no matter how.* Or it might end ". . . with aliquine and dehydration resulted" and then *however* would be a synonym of *but.* If the latter meaning is intended, then a comma after *however* effectively bars the garden path to the reader.

   To sum up: The legislation concerning the use of the comma in English sentences is in a mess. There are no firm rules, and usage differs widely even among careful writers. But would we have it any other way? The Germans have a rigid system whereby each subclause must be separated by commas. If you look at a German text, you will see, that this is so. But would you like, to copy such a rigid system? I think, that you would not.

   So let us have a round of applause for our lovely unruly flexible common-sensical language. Use both hands, please.

23 *Alphabetical Disorder*

We have all been taught sometime that the invention of the alphabet is one of the great achievements of the human mind, and that "alphabetic" writing is a form of communication much superior to "pictorial" writing. This is true, but don't let us get carried away. I shall astonish everyone by declaring that English writing and Chinese writing are not so very different, after all.

One of my most trusted advisers in matters of linguistics is a six-year-old, and he informs me that reading of English, these days, is taught by familiarizing the pupils with word outlines. Chinese reading is taught the same way. The constituents of the English word outline (called "letters") offer the reader some guidance to pronunciation. Among the constituents (called "radicals") of the Chinese word outline there is one that performs the same function. It turns out that the difference between the two writing systems is quantitative rather than qualitative: a Chinese word outline (or "character") is an arrangement of some among 214 available radicals. An English word outline is an arrangement of some among 26 letters.

And even that is a chauvinistic overstatement of the virtues of our alphabet. By pressing down the shift key of my typewriter, I add another 26 symbols to the original 26. The printer can double the new total yet again by changing from roman (upright) type to *italic;* he can treble it by switching to **boldface.** A keyboard compositor operating modern printing equipment can, in fact, generate nine alphabets (printers always use this term in the plural): upper and lower roman, upper and lower *italic,* upper and lower **boldface,** upper and lower ***bold italic,*** and SMALL CAPITALS. Nine times 26 makes 234 alphabetical symbols, and these form only one of many "fonts", i. e. families of letters so designed that they will harmonize with one another. By changing from one font to another, we may wind up with letters of a shape so wildly different that it confuses the human eye, let alone the optical scanning device of a computer.

By now, of course, my diatribe against the alphabet has gone too far. It is time for me to admit that anyone wanting to read English only has to memorize about 40 letter-shapes, and that he will soon recognize them no matter in what fanciful guise they appear. But the fact remains that an enormous variety of typefaces is at our disposal, and that we should choose among them so as to create, not confusion, but better communication.

The chemist is not often called upon to make a choice between fonts of type; he generally sends his articles to a journal or review, where the choice is made for him. But he may write a monograph, and the publisher may discuss the layout with him. Also, when his firm purchases a new typewriter, he may have a vote in the selection of a typeface. Hence it may not be amiss if I sound a word of warning: among the many fonts available, there is one group, very fashionable at the moment, which should never be used for scientific texts. This is the group of sans-serif fonts. Serifs are the little spikes or spurs which widen the beginning or end of the lines and curves that make up our letters and their abolition is demanded by the same aesthetic revolution that swept the "gingerbread" ornaments off our city buildings.

Now, I am very much at one, in spirit, with this revolution. I like austere, uncluttered architecture and simple furniture. But I only like them because the cornices and curlicues that were done away with were useless. The serifs, however, are far from useless. With unobtrusive efficiency they guide the reader's eye along the line, whereas reading sans-serif type is as difficult as scanning a graph from which the horizontal axis has been omitted. Moreover, the quest for tasteful simplicity has turned far too many sans-serif characters into look-alikes: in particular, a vertical stroke may signify the numeral one, capital I and lower-case l. Is this still communication?

I hope I have convinced you that the author of a chemical text must, at times, be prepared to argue with his layout artist. The artist's aim is to produce an attractive-looking page; generally he is very good at this. But, being more at home among mobiles than molecular orbitals, he cannot follow the chemical text and therefore does not know whether it reads well or not. He should thus not have the final choice of typeface. For an example of the mischief that can be wrought, turn to two of the most respected journals in chemical literature, *JCS Dalton* and *JCS Perkin*. Both have undergone the artist-design treatment, and in both the abstracts of all papers are set in sans-serif type. The pages look lovely; but, be honest, when was the last time you read a *Dalton* or *Perkin* abstract right through?

The chemist may not often have to choose between fonts, but quite frequently he has to choose between typefaces within the same family; he may have to make up his mind whether a certain symbol or phrase should be printed in roman, italic or bold. I shall devote the rest of this article to explaining present-day good practice. Small capitals are used by many journals to indicate valencies, e. g. iron(III), and acronyms for computer programs. Bold italic type is used for vectors. Bold roman type (indicate this by a wriggly underline) is used for matrices and, in n.m.r. spectrometry, to indicate an atom to which a signal has been assigned, e. g. $\mathbf{CH_2CH_3}$.

Far more common than these rather exotic typefaces is of course *italic* (to be indicated by a straight underline). Its name reveals its country of origin – it was first used in the printery of Aldo Manuzio, one of the great figures of the Italian renaissance. In the hands of this superb scholar and publisher printing ceased to be an imitation of ornate handwriting and became an art-form on its own. Chemists, physicists and mathematicians have special reason to be grateful to him: but for the harmonious coexistence of roman and italic on the same line scientific communication would be a very cumbersome thing indeed. Imagine, for instance, the amount of circum-scription that would be required if an n.m.r. spectroscopist could not, by a simple change of typeface, distinguish between H, the proton whose resonance he is measuring, and *H,* the magnetic field!

In algebraic expressions, italic type should be used for all those quantities on which a mathematical operation can be performed, but symbols indicating the nature of an operation (log, exp, sin, tanh . . .) are printed in roman type. We might thus have

$$dx/dy = \ln z + C$$

In chemistry, quantities that appear in mathematical equations are itali-cized, species that occur in chemical reactions are not. There are some custom-dictated exceptions, such as pH. Also, there exists a convention that by enclosing a formula in brackets we may change its meaning from "chemical species" to "quantity", i. e. concentration or activity. In nomen-clature, all stereochemical descriptors *(R, S, E, Z, cis, syn, endo, gauche)* are printed in italics, likewise "indicated hydrogen" (e. g. 9*H*-acridine). Of prefixes, only those indicating the location of named substituents are italicized (e. g. *ortho),* all others, such as cyclo, nor, seco, etc. are printed in roman. Some journals observe the custom of italicizing the names of new compounds. Note that in this instance all those parts of the name that are italicized in ordinary text are now printed in roman: thus in your manuscript these parts should not be underlined at all. (Most authors, trying to do the right thing, underline them twice, but double underlining signifies small capitals.)

Botanists and zoologists use italics for the names of genera (the first letter is always capitalized), species, subspecies and variations (first letter always lower-case, even in titles) but for nothing else; let *Gratia amazia* Myles (Editoridae) serve as an example. Writers about language italicize those words which they offer to the reader for inspection; for instance, this series of articles has often warned readers against using *using* as a preposition.

What about italicizing for emphasis? Very occasionally, when the findings of an entire paper can be summed up in one theorem-like sentence, the entire sentence deserves to be so emphasized: *Only those exemploids that give a positive Sauvequipeut test have the* cis *configuration at the junction of rings* F *and* G. But be sure the sentence deserves such emphasis before you advertise its importance so dramatically. The reader may not share your faith in the Sauvequipeut test; he may gain the impression that you are one of those who think they can win an argument by shouting. *In other words, the device does not work if it is rashly overused.* It *does* work, however, if you judiciously italicize one or two *key* words in the sentence that matters so much to you.

Finally, let us discuss "foreign" phrases; does one write *vice versa* or vice versa? The latter is correct, because the expression no longer sounds foreign. Any alien expression that gains currency in English sooner or later takes out its naturalization papers by passing from italic to ordinary type. The 1967 edition of *Hart's Rules for Compositors and Readers at the University Press, Oxford* still lists laissez-faire, rapprochement and per capita as expressions to be printed in italics; I doubt whether this will be the case in a new revised edition of the excellent work.

In chemistry, as I write, we have *ab initio, in situ* and a few other expressions. Let us not lightly add to their number. If a foreign expression is well known and has no English equivalent,* use it, *faute de mieux.* But over-

---

* Use of a foreign phrase that can be replaced by a common English term is pointless snobbery; thus *vide infra* and *vide supra* should not be part of the Chemist's English. I am also against *in vacuo* because an oversupply of italic letters in the Experimental section of chemical papers takes the emphasis away from those expressions, like *exo* and *m/z,* that depend on italicization for their effect. "In a vacuum" reads just as well.

indulgence in this habit adds to your prose a certain snobbish *je-ne-sais-quoi* which never fails to render the reader *molto furioso.*

There is also the amusing case of those words which we italicize because, chauvinistically, we want them to remain foreign. In German, *verboten* is a pretty exact synonym of *prohibited.* If we use this word in English, we give it the connotation of "prohibited by a narrow-minded, dictatorial bureaucrat". Hence the italics; perhaps it is similar disapproval of rigid rules and prejudices that leads us to italicize *faux pas, de rigueur* and, who knows, even *de facto?*

24   *Is You Is or Is You Ain't My Data?*

Do you recognize the mangled quotation? A very popular jazz number, written in 1943 by Billy Austin and Louis Jordan, was called "Is you is or is you ain't my baby?". It was a delightful piece, and I was particularly enchanted by the first line. It sparkles with Afro-American word-wit and was surely written not in ignorance but in sly disdain of the "correct" grammar of the Establishment. The tune, incidentally, goes right along with the spoof. Through most love lyrics of oppressed ethnic groups there runs a fine syncopation of self-mockery. Unrequited passion is dreadful, but not quite so bad when you compare it with potato famines, slave plantations, and pogroms.

But I digress; I was going to say that the grammar of the Establishment well deserves a certain amount of mockery. When it comes to matching verbs to nouns, English is by no means the most logical of languages. Especially when we have to choose between singular and plural, we sub-jugate grammar to common sense in a way that would horrify a German or Italian. We say "a number of compounds *were* isolated" and "kinetics *deals* with the progress of reactions". Translate these sentences literally into French or Russian and you will hear a few fine belly laughs.

I had considered writing a chapter on "nouns of multitude" such as *number,* but I find myself unable to add anything original to the brilliant articles of Fowler/Gowers (pp. 260, 402, 403 of *Modern English Usage*) and Bernstein (pp. 108, 221 of *The Careful Writer*). I shall instead concen-trate on another singular–plural problem that gives the chemist much bother: those nouns of foreign origin which may or may not take a foreign plural form *(formulae? formulas?)* and those whose status as singular or plural is uncertain (data *is?* data *are?*).

This is a problem we have created for ourselves. When a foreigner enters a community, is well received there, and makes his home on it, he is expected, after a few years, to take out citizenship papers and obey the law of the land without relying on his consul for further protection. It is we, in our snobbish folly, who have given some of the newcomers privileged status. We say *spectra* rather than *spectrums* because we fear the latter term might give offense to the Latin embassy. We would be quite justified, and it would make good linguistic sense, to engage in a Boxer Rebellion and bid all foreign devils to cease such gunboat diplomacy. But custom and snobbery make slaves of us all, so I shall describe the situation as it is rather than as it ought to be.

Exempt from the preceding tirade are those foreigners who only arrive for brief visits and have no intention of colonizing the natives. For such words, the foreign form of the plural adds to their exotic flavour. If you know more than one *lingua franca,* your fine scholarship entitles you to refer to them as *lingue franche.* No one could accuse you of the ostentation one so deplores among *nouveaux riches.* The plural of kibbutz is *kibbutzim;* and *bedouin* is a plural term that does not need a final *s.* Japanese nouns remain unchanged in the plural, and thus it is quite correct to say: *Seven samurai were attended by fourteen geisha.* Good luck to them; we must regretfully tear ourselves away from the outlandish spectacle and return to the familiar fields of chemistry.

Let us begin with an easy word. *Spectrum* is the singular; *spectra* is the plural. The quite permissible *spectrums* is unknown in scientific literature. Nearly everybody gets this word right, but during the past few years I have edited manuscripts in which *spectra* has been mistaken for the singular. It is quite possible that, in a few decades, the troubles that now beset *data* will bedevil *spectra* also.

Data is the plural form of the Latin *datum,* "something given". These days it is used as a near-synonym of *results* or *findings* (except that *data* more specifically refers to numerical values). This is a pity; originally *data* was meant to be the information at hand at the beginning of an investigation, *results* that at its completion. *Datum* was so little used in English writing that eventually *data* was mistaken for the missing singular. At the moment, it seems that this mistaken use is going to prevail; in most newspapers and scientific journals "the data is" and "the data are" can be found side by side. However, Fowler and Gowers, and Bernstein, still denounce "the data is" as a solecism, and the *New York Times* as well as most respectable book publishers still insist on treating *data* as a plural form. So does the *Australian Journal of Chemistry;* we have no great illusions that we shall win the battle, but at least we are fighting it in the only way it can be won, i. e., by reinstating the unjustly suppressed singular *datum.*

Agenda is a word for which the battle has been lost. Originally it was the plural of *agendum,* but that singular never entered the English language. *Agenda* is now commonly considered the singular, and occasionally one encounters the new plural *agendas.* Bernstein explains the difference: "*Agenda* has departed from its original meaning of things to be done, and now means a program of things to be done . . . *Data,* on the other hand, has retained its meaning of things or facts."

Formulae and *formulas* are, at the moment, being used equally frequently. Most editors allow both plurals to coexist, but it is only a matter of time before the Latin plural will disappear.

Index rejoices in the possession of two plurals having distinct meanings. The Latin plural is used when the word has its mathematical sense: exponential *indices.* In all other senses (author *indexes,* prosperity *indexes*), the English plural is used and will no doubt soon displace the Latin one from its last stronghold. Latin gunboat diplomacy is rapidly collapsing as far as words ending in *-ex* and *-ix* are concerned. *Appendixes* is used almost universally, and the Latin forms *vortices, apices,* and *cortices* are disappearing. The greatest life expectancy is that of *matrices.*

Greek gunboat diplomacy is also having its ups and downs. It is completely effective with *phenomenon,* which everyone multiplies to *phenomena. Criterion* is nearly always given the Greek plural *criteria,* but here a *data*-like situation is developing in that some writers have begun to mistake the plural for the singular. There is no need to tell the readers of this book that all subatomic particles such as *protons* take only the English plural. There would be a nuclear explosion of considerable intensity of someone tried to talk of slow *neutra!*

This brings us back to Latin: Why do we, unanimously, say *nuclei* when we are about evenly divided between *bacilli* and *bacilluses?* If you encounter a Latin word ending in *-us,* do not try to form the Latin plural unless you still remember your Cicero and Tacitus well. Not all of them take the plural ending *-i;* some *(apparatus, status, prospectus)* remain unchanged and others vary in unpredictable ways (*corpus* turns to *corpora*). You will always be safe with an English plural.

Most of the Latin words ending in *-um* have remained safely in the shadow of the gunboats. All chemists without exception talk of *maxima* and *minima,* nearly all biologists of *bacteria.* We have already talked of the curious case of the disappearing singular in the case of *data* and *agenda.* Another word that may soon be in a similar situation is *media;* its singular is disappearing from circulation and the first semiliterate fools are already beginning to mistake the plural for it. Perhaps the best remedial action for *media* is to suppress the word, which by now must be getting on everyone's nerves.

A word about modern languages. I remember insisting on *plateaux* in the first paper on which I collaborated; but that was long ago, and these days I write *plateaus* and *bureaus*. It is a sign of the times that few German words are left in the chemist's English. We treat *zwitterions* (quite appropriately, in view of its meaning) as a German + English plural. In a similar manner, we have converted *Eigenwerte* into *eigenvalues* and a number of other *eigen*words have English eigenplurals.

This completes our *conspectus* (plural *conspectuses* − *conspecti* would make you a laughing-stock among Latinists) of foreign words with deviant plural forms. I have left myself space to deal with one more type of solecism, the one Fowler calls the "red-herring plural". A writer chooses as the subject in his sentence a noun which is in the plural. Before he can get to the verb, he encounters another noun which is in the singular, and absent-mindedly he matches his verb to that: "The results, obtained in an investigation which predates the appearance of Tizio's paper, *shows* that ..." Do you shake your head pityingly? Of all the grammatical errors I encounter in the manuscripts I edit, this is the most frequent − I have counted up to five such lapses in a 10-page paper. Now surely there is no need for me to preach a lengthy sermon on the subject that *they is* and *he are* are not good English. You know this as well as I do; these frequent errors are not the result of ignorance but of distraction. Your best protection against embarrassment of this nature is to get yourself a critical, independent-minded assistant and ask him or her to read your draft carefully. More harm has been done to science by overrespectful collaborators who admiringly confirm every word the Herr Professor utters than by any other cause. And if you want to argue this point with me, think again; after all, I have the *data*.

25 Yes, Virginia, There IS a Temperature

He who writes about language skates on very thin ice indeed. One moment I was gaily pirouetting on the surface on things, tracing intricate mathematical symbols; the next moment there was that telltale crack and I found myself in the icy waters of philosophical controversy. Give me a seat by the fire and let me tell you what happened.

It all began with a purely typographical dispute about the way column headings of tables in scientific papers should be set out. You agree, I trust, that tables form part of The Chemist's English? They are just a shorthand way of writing down a series of nearly identical English sentences. Say you have just determined the thingummicity of examplamine and 48 other substances. Rather than repeat ad nauseam: The thingummicity of examplamine is 72.4; the thingummicity of paradigmol is 56.3; ... you arrange your substances in one column, your numerical values in another. Over the first column you write the heading "Compounds" and over the second "Thingummicity". These column headings ("box heads", the printer calls them) present no logical difficulty. But then thingummicity, as everyone knows, is a dimensionless quantity. Suppose you are dealing instead with entropy, which is measured in $J\,K^{-1}\,mol^{-1}$. Naturally you will not want to write these units, tediously and with useless consumption of space, beside every numerical value. They, too, demand to be included in the box head. Thus the heading of our column must contain the symbol for entropy, S, followed by the units in which it is measured. The question, at first, was simply how to keep these various symbols from getting mixed up.

Here I must admit that, much to the discredit of the editorial profession, different journals adopted different solutions. Some relied on the fact that symbols for physical quantities are by convention printed in italics, symbols of units in ordinary ("roman") type, and printed all symbols side by side. Others placed a comma after the "quantity" symbol. The system still preferred by the *Australian Journal of Chemistry* is first to print the "quantity" symbol and then, *separated by a space* (please make a careful mental note of this in view of what follows), the units in parenthesis; thus

$$S \text{ (J K}^{-1}\text{ mol}^{-1}) \qquad \text{(note the spacing)}$$

Naturally, most editors prefer to print the symbol for the physical quantity on one line and that for the units on another, but this is not always possible, and such a layout disguises the problem rather than solves it. Moreover, the numerical entries in a column often have a multiplier such as 10^6 in common; if this is to be transferred to the box head, where should it appear?

I am glad to report that this last problem was solved, with characteristic resolute commonsense, by R. S. Cahn. The Chemistry Society "Handbook" insists that the multiplier be made to *precede* the symbol of the physical quantity; thus an entry 3.6 under a box head 10^6C indicates that the value of C is 3.6×10^{-6}. The *Australian Journal of Chemistry* embraces this notation with fervour; it is the one certain protection against confusion.

But the confusion about the separation of the "quantity" symbols persisted. An attempt to clear it up was undertaken in the 1950s by Professor E. A. Guggenheim; after a lucid critique of all common notations he advocated the use of one system, the so-called "quantity calculus", as being logical and consistent; and indeed the system has now been taken up by IUPAC. I honour Guggenheim's memory, and I am aware that I shall have to spend the rest of my days in hiding from a lynching party of thermodynamicists if I criticize his notation. Nevertheless, there are some flaws in it, and it is time that someone brought this matter into the open.

Quantity calculus (Guggenheim attributes the system to A. Lodge) bravely sets out to define a term that is vague in most scientists' minds, the term *physical quantity:* "A physical quantity is the product of a pure number and a unit". Thus the sentence that, in the English language, reads:

$$\text{The temperature is 393 kelvin} \qquad (1)$$

becomes "temperature equals 393 times kelvin" and can be written, in the language of mathematics,

$$T = 393 \text{ K} \qquad (2)$$

The unit K can of course be brought to the left-hand side of the equation, which then becomes $T/\text{K} = 393$, and indeed this is the notation IUPAC recommends. To give two more examples, an entropy and a first-order rate constant:

$$S/J\,K^{-1}mol^{-1} = 15 \tag{3}$$

$$k/s^{-1} = 0.3 \tag{4}$$

Quantity calculus has some very great virtues: it is an effective teaching device, and it can be a great help in coping with inconsistent systems of units. But it is a fatal error to bring too rigid a logic to bear on what is essentially a problem of language. The notation purports to effect a neat translation from everyday English to the language of mathematics; instead, it serves up an unhappy pidgin. It is a tenet of the grammar of mathematics that *a full space to the right of a symbol signifies the end of an operation.* Look at expressions (2) and (3) and you will note that they commit the solecism of making the space serve the function of a multiplication sign. Recent manuals have tried to remedy this error in logic by inserting multiplication signs between the symbols of units (but not between numeral and unit). But this causes the notation to lose all value as an editorial device. How can such a cumbersome term as

$$S/J \times K^{-1} \times mol^{-1}$$

be fitted into the "box head" of a table, or divided at the end of the line? (I ask my readers to believe me that the problem will not go away if the "slash" is replaced by a horizontal "fraction rule"; the reasons are typographical and would take too long to explain.)

Moreover, we have been taught in school that the expression a/b^{-1} can be replaced by ab; but if we were to perform this operation on equation (4) would we obtain an acceptable notation? Finally, in most papers there occur physical quantities that are not expressed by symbols but spelled out. I for one cannot lightly bring myself to accept a column heading such as "reaction time/s".

So far we have only talked of superficial matters, but there are more profound problems to come. The advent of quantity calculus has subtly altered the relation between the language of mathematics and that of everyday life. Take the expression $f = ma$; this is commonly "translated" into English as *force equals mass times acceleration.* Before the arrival of the new notation, such a sentence could only be described as a sloppy shorthand for "In a self-consistent set of units, the numerical value of the force is equal to the product of that of mass and that of acceleration". With the new definition of physical quantity, the italicized expression above becomes rigidly correct.

But we pay a very heavy price for this gain. If it is now strictly correct to say "force equals mass times acceleration", it becomes strictly incorrect to say that a quantity *is proportional to the logarithm of the pressure.* Indeed, only pure numbers have logarithms; but the newly defined pressure is not a pure number but the product of one with a unit. This is no mere hairsplitting; the problem is real. Chemical papers abound with expressions such as $\log k$ and $\exp(R/T)$; the new notation reduces all these to nonsense. Organic chemists — a pragmatic tribe with no particular interest in linguistic philosophy — will not take kindly to the suggestion that their familiar $\log \varepsilon$ would now have to be written, if you please:

$$\log(\varepsilon/\mathrm{l\,mol^{-1}\,cm^{-1}})$$

Not so long ago, I received a paper which contained the expression "$\exp(2.6/T)$" where T was meant to symbolize the temperature. The referee pointed out that this was dimensionally incorrect and should be amended by inserting the sign K for kelvin, i.e. $\exp(2.6\,\mathrm{K}/T)$. But the paper was about potassium! The author and editor took a prudent retreat and defined T not as the temperature but as the *number* of kelvin.

Now please do not misunderstand me. I am not an implacable opponent of quantity calculus. The journal I help edit permits its use, except where symbols of units suddenly invade the "heart" of an equation and intermingle with numerals and quantity symbols. But we should all remember that "quantity calculus" is not an immaculately perfect system, a revelation sent from heaven to heal the breach between the English and mathematical languages. It is a convention like other conventions, and the first requirement of a convention is that it be convenient. On those occasions when it is not convenient, it should not be used.

It was at this moment that I heard the ice crack. I had been looking for proof that the new notation was not quite as logical as it sounded, and found myself comparing the expressions:

$$T = 393\,\mathrm{K} \tag{2}$$

$$(a + b)^2 = a^2 + 2ab + b^2 \tag{5}$$

Now I submit that the two = signs are not semantically equivalent. In equation (5) the equality sign expresses an immutable verity, destined to endure as long as the language of mathematics endures. In equation (2) it expresses an ephemeral equivalence which is at the mercy of time and place, of the accuracy of the thermometer and the alertness of the observer. The equality sign of equation (2) is simply a very approximate translation into mathematical language of the verb *is* in our sentence (1). And this *is* is far closer in meaning to the verb *is* in

The temperature *is* unbearable   (6)

which cannot be translated into mathematics at all, than to the sign in equation (5).

It is from this first imprecision that all further difficulties flow. The two languages, English and mathematics, do not cover identical territory. There are thoughts that can be expressed in one language but not in the other, and it is wrong to assume that mathematics is all-embracing and to allow it to dominate communications. At some stage in our scientific work, we all have to switch from English to maths. We heave a deep sigh, take the value we have just read off our carefully but not infallibly calibrated instrument, *decide to treat it as a pure number,* and insert it into the appropriate equation. But we should be aware that we are carrying out an approximate translation, with a subsequent gain in $S/\text{J K}^{-1}\text{mol}^{-1}$.

By now we are in very cold philosophical waters indeed. We have drifted into ontology, that branch of philosophy that attempts to define the word *to be.* What do we mean when we say the temperature *is* unbearable, *is* 393 K? Does it make sense when we say *there* is *a temperature?* Do we all mean the same thing by *is?* Think back to the exchange of letters known as "Yes, Virginia, there is a Santa Claus" and note how the partners in the dialogue talk past each other. For the enquiring child the criterion of *being* is the possession of a set of (white-whiskered) space—time coordinates, proximal to those of chimneys and red-nosed reindeer. For the journalist composing his homily, who treats Santa Claus as a synonym of generosity, the criterion for *being* is the property of definable-ness, of having perceivable limits. There *is* such a thing as generosity because we can perceive the boundary between generosity and greed.

The search for the meaning of *is* seems such a pointless pursuit, and yet battles have been fought for want of this knowledge. Men have been burnt at the stake for affirming or denying the proposition: "There *is* such and such a deity". Yet neither the performer of the oxidation nor its substrate could have defined the word *is* well enough to make sure the disagreement between them was real.

The problem does not seem to have bothered Western philosophers much until, I think, the late Middle Ages. But in China, it was very clear to the Taoist sages two millenia ago. That limitless, undefinable thing, the Tao, is in the writings of Chuang-Tse often referred to as "that which is and which is not".

Which proves one thing: communication is a tricky business. A paradox, which cocks a cheerful snook at logic, is often more efficient at getting your meaning across than a painstakingly constructed syllogism. Thermodynamicists and similar theologians please note.

At this point I shall bring our discussion, which has drifted far away from typography, to a hasty close. Otherwise gentlemen of Strong Religious Convictions will start writing impassioned letters to the publisher. Which Zeus forbid.

26 *The Truth about the Truth*

I looked at the clock and saw it was High Noon. With the last gulp of rye burning in my throat, I rose from my bar stool, gave the barmaid a crooked smile, spat the cigarette stub into the spittoon, pushed the swinging doors of the saloon open and walked out into the sunbaked streets of Fort Guggenheim.

Now, as the camera gets weary of following me past cowering dogs, past Mexican women scurrying into alleyways after one compassionate glance, and past all the other visual clichés that are unfailing harbingers of the approaching gunbattle, let us have a flashback. In the preceding Chapter of this series, I had spoken about the IUPAC-favored practice of "quantity calculus" in such terms that it seemed my only hope of avoiding the vengeance of outraged thermodynamicists would lie in growing a beard, changing my name, and going to live in Broome, in remote north-western Australia. And indeed I had begun to pack my bags, when the thought struck home that life far from the bright city lights and from *Chemical Abstracts* was not worth living. And so, with weary courage, I decided not to flee but rather to seek a showdown, and mailed copies of my typescript to some prominent Australian scientists who I thought would have strong feelings in the matter.

Observe, as our stereophonic soundtrack now picks up the hiss of the first bullets, that the fire seems to come from all sides. I took the wise precaution of sending my screed not only to thermodynamicists, who are the most passionate practitioners of quantity calculus, but also to other theorists, to mathematicians, and to a philosopher. They all reached for their guns, but the fact that the shots came from various directions now confers on me a certain measure of safety – the fire from one side pins down those that would advance upon me from the other.

Some bullets do strike home. Professor R. H. Stokes, at the end of a helpful and stimulating letter, confronts me with a problem that I had quite inexcusably ignored. He writes: "From a teacher's point of view, perhaps the greatest weakness about the current rules is the one that says physical quantities are to be represented by italic symbols, and units by roman ones. This is all very well in printed material, but cannot be observed on a blackboard."

How easily could the confusion Professor Stokes deplores have been avoided if our founding fathers had reserved lower-case letters for quantities and capitals for the initial letters of units! But this particular horse has bolted. Observe, though, that Professor Stokes has furnished me with further ammunition for my main contention that *letters designating units should not appear in equations.* As long as we can distinguish between roman and italic, we may say that italic letters written close-up signify a product of symbols, roman letters written close-up are to be read as one symbol. If the distinction disappears, then a teacher who expresses a pressure P in pascals and writes, in the IUPAC notation, P/Pa = 5 may find that a student has cancelled the P from the left-hand side of the equation.

Thermodynamicists will contend that this is just a semanticist's punctilio, and no student would ever be that stupid. And of course it would be foolish to deny the great virtues of their notation. To quote Professor Stokes again: "Instead of

$$\ln p = 21.5 + 310/T$$

which has no meaning unless you separately specify the units of p and T, you write

$$\ln(p/\text{atm}) = 21.5 + 310 \, \text{K}/T$$

This is a small price to pay for complete clarity, and a statement which cannot be misunderstood 50 years hence."

Other leading workers in this field, in particular Dr S. D. Hamann, have written in similar terms. But, as I dodge this hail of bullets, a close-up shows my face to be unrepentant. Let me fire a few shots back: (1) Where I stand, the "price" can be expressed in dollars and cents, and can be very large. (2) Any equation that is too lavishly festooned with symbols is apt to be mis-edited by the editor, mis-printed by the printer, and mis-read by the reader. (3) In Chapter 30, I shall prove that the symbol K has at least four meanings in current literature. Thus, although I am as passionately concerned as Stokes and Hamann that the units of all symbols sould be defined, I believe such definition should occur just before or just after the equation.

In supporting this view, I have found a powerful ally in Professor D. P. Craig. He writes: "Referring to those naughty expressions such as log k, I think of them as shorthand for the logarithm of a ratio, namely log (k/k_0), where k_0 equals 1 in the same units as those used for k. I can convince myself this is an acceptable shorthand."

I entirely agree, and so do my colleagues on the *Australian Journal of Chemistry*. A suggestion similar to Professor Craig's was also made by Dr T. H. Spurling.

A stray bullet has just caused a flesh wound. I had written that organic chemists would not take kindly to the suggestion that log ε would have to be written as log $(\varepsilon/1 \text{ mol}^{-1} \text{ cm}^{-1})$. "Organic chemists" writes Dr Hamann "will have to learn. Britain *et al.* ("Introduction to Molecular Spectroscopy", Academic Press 1970) list values of ε in "the SI units" $\text{m}^2 \text{ mol}^{-1}$, which are a factor of 10 greater than $1 \text{ mol}^{-1} \text{ cm}^{-1}$. This may well start a fashion." Dr Hamann has taught me a valuable lesson and I ask all authors to specify their units of ε – but not, I implore you, in equations and the "box heads" of Tables.

Dense gunsmoke now enshrouds the combatants as the philosophical battle commences. I had submitted that in the two expressions

$$T = 393\,\text{K} \tag{2}$$

$$(a + b)^2 = a^2 + 2ab + b^2 \tag{5}$$

the = signs were not semantically equivalent. This brought forth volley upon volley of brilliant correspondence. Professor Craig again comes to my help by professing to share my distrust of the "equals" sign in (2). "Your expression (2)" he writes "would be acceptable to me only if the value 393 K had been found by solving an equation for the variable T". But he then joins forces with Professor Stokes and Drs J. E. Lane and T. H. Spurling in taking me to task for blurring the distinction between two concepts. Expression (2), say the members of this posse in unison, is an equation whereas (5) is an identity. In Professor Stokes's words "Mathematicians recognize that there are two different uses of the equals sign, and sometimes get round the problem by writing the identity sign [the triple bar] to mean 'is identically equal to'."

To this I reply, emerging briefly from behind a verandah post, that I have not created the confussion, I have merely pointed to its existence. We only use, in scientific texts, the triple-bar sign in the sense "can be more conveniently written as". Otherwise we use the double-bar sign in the sense of "can be transformed into". And thus I remain troubled over expression (2).

Dr Hamann has just boldly stepped between the fighters by declaring that no grounds for dispute exist. He tells me "I'm afraid I can't see your distinction between the = of (2) and of (5). To me, they can both be translated perfectly well into English by the dictionary rendering 'is neither less nor greater than'."

While the attention of the posse is thus diverted, I attempt to make a dash for it. But my path is blocked by a philosopher. Professor J. J. C. Smart writes: "I think you have mixed up two unrelated issues. Expression (2) is empirically true, (5) is mathematically true; but this does not affect the fact that = means the same in both cases. 'The professor of anatomy = the Dean of the medical school' is empirically true; '2 + 2 = 4' is arithmetically true. But in both cases = means 'is identical with'. It is an unfortunate accident that mathematicians have come to say *equals* for *is identical with*. The symbol 2 + 2 is not the same as the symbol 4, but they mean the same object, the number four. Confusion of difference of symbol with difference of object symbolized makes people think that '2 + 2 = 4' says that *two* things are equal, like two professors in a department. In logic = is read 'is identical with' and when mathematics is formalized as in *Principia Mathematica* it becomes quite clear that = is the symbol of identity."

Here is another extract from Professor Smart's letter: "'T (in K) = 393' is just to say that the real number 'T in K' is the very same real number as 393. I don't think that T = 393 K is analogous to 'the temperature *is* unbearable'. In the latter the *is* is the *is* of predication. (Perhaps one day language will be more like logic and we shall say 'the temperature unbearables' or 'the rose reds'.)"

As an amateur linguist, I am enormously grateful to Professor Smart for his final statement; it illuminates a problem of language I had never properly understood. As for the rest of his impressive analysis, is it not about time we chemists closed ranks and attempted to discover in what way our language differs from that of the philosophers? Certainly their precise logic reduces all our presumed "equalities" to "identities". But is their rigid and severe language good enough to supply our needs of communication? Do we not, besides the all-embracing term *identity,* need another word signifying *transient identity,* such as occurs in the measurement of a temperature? And can we really be satisfied if, after having subjected a dozen tricky parameters to a complex series of mathematical operations and emerged with a satisfactorily symmetrical expression, we are told that we have not really discovered anything new but just re-stated an identity? If we must use the term *identity,* then surely it is fair to say that we use the = sign in the sense of *identity revealed.*

Philosophers and physical scientists are often called seekers after truth, but their aims are not quite identical, and this has led to the present difficulty of communication. The philosopher is interested in the *true* statement; i.e. one that is immune from contradiction. The physical scientist aims at the *valid* statement; i.e. the one that has the greatest power of prediction. Thus, if we say that "examplamine may react with McInstance's reagent or it may not" then this statement is absolutely true but also very shallow. But if you affirm that "examplamine will react with McInstance's reagent to form paradigmol" then your statement may only be a half-truth but still be scientifically very valuable. The scientist, truth to tell, is not comfortable with the word *truth.* If someone were to ask us whether Boyle's law was true we should certainly refuse to say yes, but we would be equally unwilling to say no.

While you ponder whether the truth of these statements is valid, I shall quote again from the letter by Drs Lane and Spurling: "Because a convention has been called quantity calculus it has been mistaken for mathematics. It is not mathematics but a scheme of identifying the dimensions of the unit part of a physical property. If this is accepted, the problem of spaces between symbols disappears and the string of characters $J K^{-1} mol^{-1}$ (including blanks) represents a single unit." This is sensible talk and greatly appeals to the chemist in me. The editor in me, though, is still unhappy because the "single unit" still occupies eleven spaces.

Land sakes, them comma rustlers can shore talk up a storm. By now dust swirls everywhere and the gunbattle is called off. The hero, bloody but unbowed, limps towards the saloon, to be bandaged with the heroine's petticoat. As the final credits roll over the closing sequence, I am discovered with my arm around the barmaid's shoulder as we both read, with breathless stimulation, the latest issue of the *Journal for Chemical Information and Computer Sciences,* just arrived by stage coach. These modern philosophical westerns are a lot sexier than the oldfashioned kind, you know.

27 They Also Serve who Only Pull and Tug

Consider, o reader, the noble ocean liner. Proudly she glides on her way; haughtily her prow stares down on waves that would have overtopped the frail craft of earlier ages. The whale, giant of the sea, looks at her once and flees. The shark, king of the waterways, becomes a beggar for her sake and follows her about, greedily asking for the scraps from her table. The albatross, whose wingspan awed the ancient mariners, might perch in one of her numerous lifeboats, and not be noticed. The dolphins that gambol in her bow-wave are like pageboys holding up an empress's train.

But once she arrives in the harbour, how quickly and shamefully does our empress lose her dignity! The narrow confines of pier and wharf concede no room for graceful movement. She who was queen of the sea now has to engage in humiliating traffic with those fixers and hustlers and blackmarketeers of the port, the tugboats. All majesty gone, she is pushed and tugged and pulled along at the end of hawsers. The epauletted captain on her bridge now seems nothing more than the stooge of the unshaven tobacco-chewing veteran below who turns the wheel of his grimy little craft.

The tugboats of language are called auxiliary verbs. You have, perhaps, an important thought to express, and want to express it as vividly as possible. Out of the wine-dark recesses of the mind a noble verb swims into your ken, but before can berth it in your sentence much tooting and pulling is required. Open any book at random and you will notice how very few sentences on the page would survive a strike of tugboat captains.

English is the Tugboat Annie of languages. More than any other verb in the western world, the English verb is dependent on its auxiliary verb to convey shadings of tense and mood. There are no less than nine manoeuvres that the French verb can execute under its own steam. The English verb is capable of only two: the expression of present and past. Perhaps some reader, clutching a grammar primer, may object that the English verb is also able to express, unaided, the subjunctive. But I doubt whether he *consider* himself capable of supporting this argument from examples of contemporary Australian speech and writing. (The subjunctive is still showing signs of life, *be* they ever so tenuous, in the writing of educated Americans, much to their credit.) The English verb is still able to reach, self-propelled, the berth of another verb form, the imperative; but for anything beyond that, *bring on* the tugboats!

27.1 Will the Past Be Abolished?

The late Lin Yu-Tang once argued, in a witty article, that English had now reached the stage Chinese occupied 10,000 years ago. Certainly what I have said above supports his argument: the English verb is rapidly approaching the state of the Chinese verb, which has neither number nor tense. I must even qualify my statement that the English verb can still express, unaided, the present and past tense. This is no longer true if our sentence is a question, or in the negative. Milton's England *lacked not* speakers who would use a direct negative but, these days, *know you* of any? (It is true that President Kennedy, in a mood of exaltation, once said *Ask not what your country can do for you.* But that is the sort of thing even the most powerful ruler can get away with only once.)

The past tense, meanwhile, has troubles of its own. Increasingly it is being supplanted by the tugboat-attended present perfect. The modern chemist is apt to say *we have found* in instances where his Victorian predecessor would have said *we found.* This is a danger against which we should be on our guard; the distinction between the two tenses is a valuable asset of our language.

The constant traffic of tugboats may seem an undignified spectacle. Indeed modern English is the least majestic of languages, a hostile harbour indeed for the full-blown orator – whenever we encounter a sequence of rolling phrases, we immediately suspect a leg-pull. But what our tugboats lack in dignity they make up in efficiency. With faultless accuracy they guide our verbs towards the exact shade of meaning that we had intended. Conversational phrases such as *we may come* and *we might come* convey varying degrees of likelihood and enthusiasm with a precision no other language I know can match.

So let us look at our tugboats with new respect as, sometimes singly and sometimes in pairs (*we should have* come), they guide our glistening verbs to their appointed berths. But while we admire their teamwork, let us not forget that it is not completely faultless. There are some curious union rules and demarkation disputes on our waterfront. The tug that takes your verb to its wharf is not always the same that pulls it out again. For instance, the negation of *we may assume* is not *we may not* but *we need not assume.* This is a custom hallowed by long popular usage. A far less harmless practice is the inversion of functions of *shall* and *will.*

27.2   *I Shall. Wilt Thou?*

We were all taught in school that *I shall* indicates a dispassionate comtemplation of the future and *I will* a determination to bring it about. In the second and third person, our two tugboats suddenly change jobs. Fowler considers this an ancient and lovely custom well worth preserving. Dr Onions argues that faultless observance of this rule sets the native Englishman apart from such lesser breeds as Scots, Irish and Americans; Australians were no doubt beneath the good doctor's notice. English schoolchildren are taught the story of the Scot who fell into the Thames and called out in despair: "I *will* droon and nae one *shall* save me." The assembled Englishmen, all avid readers of Onions's *Advanced English Syntax,* respected his apparent death wish and left him to his watery grave.

As an editor I must naturally enforce the dicta of these eminent authorities. As a private person, I respect their views as little in theory as the vast majority of English-speaking people respect them in practice. Of all the useless shibboleths foisted by clerkish snobbishness upon the language, this seems to me the worst; its social value is equivalent to that of daintily extending one's little finger during tea-sipping. So, the next time you find yourself saying *I will* when your schoolmaster would have said *I shall,* do not blush; you have done nothing wrong (except if you find yourself in the Thames and cannot swim).

27.3   Double-Duty Tugboats

One of the most generous donors of ideas to these articles is my colleague Susan Ingham, of the *Australian Journal of Zoology.* On reading the above paragraph, she disagreed and offered me, for the first time in my life, a logical explanation of the switch from *shall* to *will.* The verb *shall* (German *soll*) denotes obligation, *will* denotes mere volition. Our future is made up of the deeds we perceive ourselves obliged to do. But our neighbors' future deeds, alas, will not coincide with what we consider their obligations. We *shall* do, but our neighbors *will* not necessarily do, what we think must be done. Hence the change in auxiliary verb. I find this explanation fascinating, but not convincing; the logic spins altogether too fine a web to fish any Scot out of the river.

There are two types of tugboat traffic with which the chemist is not familiar. Is a single auxiliary verb entitled to take two main verbs in tow? If its towing equipment (i. e. its number) fits both, there is no problem:

> The solution was concentrated, and the precipitate (was)
> removed   (1)

Here, clearly, the second *was* can be eliminated. But what if one of the auxiliary verbs should be in the singular, and the other in the plural? Let us alter one word in our sentence:

> The solution was concentrated, and the crystals (?were)
> removed   (2)

Here, by omitting *were,* we are in fact saying that *the crystals was removed,* and that are poor English, if you knows what I mean. Nevertheless, this type of construction comes naturally enough to us in everyday speech. In an article headed *Ellipsis* (the technical term for the omission of words the reader's or listener's mind will automatically supply) Fowler gives qualified approval to the use of such double-duty tugboats. He objects to them only when the second half of the sentence is complex:

> The solution was concentrated, and the crystals, which had a
> bluish tinge, *were* removed   (3)

R. S. Cahn, in the Chemical Society "Handbook", takes a more fastidious stand, and advocates the retention of *were* in sentence (2). I follow his lead, partly out of respect for him and partly because I do not wish to hasten Lin Yu-Tang's prophecy of a number-less and tense-less English verb towards its fulfilment.

27.4   Triple-Duty Tugboats

The chemist's second problem concerns the use of triple-duty tugboats. If one subject governs three verbs, then the auxiliary verb tugging them need only be used once:

The solution was heated, filtered and concentrated   (4)

This causes no difficulty. None is encountered, either, when we have three nouns in succession governing three verbs.

The solution was cooled, the precipitate removed and the
filtrate concentrated   (5)

The correctness of sentence (5) follows logically from that of its simpler analogue, sentence (1).* But, in the Experimental sections of papers on preparative chemistry, one frequently finds the following hybrid of sentences (4) and (5):

The solution was cooled, the precipitate removed and dried   (6)

Perhaps I am unnaturally fussy, but I cannot accept such a construction. The reader is first led to believe that there will be a noun for every verb, and this expectation is suddenly deceived in the last two words. If this it not changing horses in mid-stream, it is at least changing hawsers in mid-harbour. A second *and* would not be amiss:

The solution was cooled, and the precipitate removed and dried   (7)

The fault in sentence (6) is trivial, I admit; all the more so as such sentences nearly always occur in the Experimental section of a paper, where no one looks for perfect prose. I amend these sentences wherever I can, but suffer no pangs of conscience if I find I have missed one. But the

---

* For further adventures of sentence (5), see Chapter 33.

carelessness of language that sentence (6) typifies is by no means trivial. If we want our traffic of ideas to continue smoothly, we must, first, make certain that the number of tugboats does not grow unduly; second, we must pay proper respect to those tugboats we have come to rely on. If we do not take due care we may find ourselves where tugboats have no business to be, namely in deep waters.

28 On the Training of Old Dogs for Which-Hunting

The psychological literature is well supplied with articles about the difficulty of teaching an old dog new tricks. This difficulty is undeniable, but how is it to be explained? Are we simply dealing with the superficial problem that, as eyesight and hearing deteriorate, less input penetrates to the brain? Or does the aging dog's ability to process input deteriorate too? Again, could not the brain's diminishing storage capacity cause the trouble?

It is hard enough to sort these three physiological causes out, but now we must also consider the old dog's motivation. The subject of our enquiry is, by now, at the height of his canine career. He has found his social niche: for years he has been regularly employed as guardian, rat-catcher and entertainer, and he believes his superannuation rights to be secure. Gone are the days when, with tail-wagging enthusiasm, he frantically strove to demonstrate his indispensability to his master. The performance of new tricks may necessitate painful and undignified contortions; and experience has taught our dog that the rewards are no greater than what ten minutes of skilful begging would produce anyway. And so the old dog may refuse to learn the new trick, not because he is not smart enough, but because he has developed a different kind of smartness.

This kind of smartness, which does credit to neither man nor beast, is occasionally in evidence in manuscripts submitted to scientific journals. Why is it, asks the editor gently, that Part LII is still full of the infelicities that gave the referee of Part XLIX high blood pressure? And back comes an ingratiating reply: Very sorry, but old habits die hard . . .

To shame such authors I shall now boastfully recount how I, over the last year or so, have partly overcome a weakness of my own in the use of the English language. I used to suffer from a certain tone-deafness for the distinction between *that* and *which*. Lately, though, the deafness has begun to lift, and for this I give thanks to the teachers who were so patient with this old dog.

Mind you, I have excellent teachers. My series of articles has been read by many distinguished editors overseas, and whenever I slip from grace I receive a number of concerned and amiable letters. (Experienced editors are almost invariably amiable people; any editor who lacks that virtue will soon be eaten by either his authors or his ulcers.) The admonitions *that* caused me to think more deeply about the difference between *that* and *which* came mostly from Dr W. E. Cohn of the Office of Biochemical Nomenclature, National Research Council, USA. In his honour, I shall open the discussion with a quotation from the CBE (Council of Biology Editors) style manual:

> "Precise usage favors *that* to introduce a restrictive (defining, limiting) clause and *which* to introduce a non-restrictive (non-defining, descriptive) clause. Maintaining the distinction between these relative pronouns contributes to clarity and ready understanding."

Amen to that. The difficulty, though, is that the above passage is written in grammarian's language. It is not easy for the chemist to see why a "descriptive" clause should not also be a "defining" one. We have to begin (my philosopher friends won't like this sentence) by defining a "defining clause".

Well, a "defining clause" is one that wraps its contents in a bag and puts an "exclusive" tag on it. Here are two examples in which defining and non-defining clauses are contrasted:

> Compounds *that* contain nitrogen play an important part in
> plant metabolism   (1)

> Alkaloids, *which* contain nitrogen, play an important part in
> plant metabolism   (2)

Sentence (1) wraps a certain group of compounds in a bag. If they contain nitrogen, they are in, if not, they are out. Sentence (2) does no such wrapping up. It does not separate off alkaloids-containing-nitrogen from alkaloids-not-containing-nitrogen; the latter do not exist.

It follows that *which*-clauses always express a property common to all members of the set of things mentioned in the main clause, and that property does not confer an "exclusive" tag. The distinction between *that* and *which* is that the *that*-clause splits off a subset from a set. Here is another pair of examples:

> The fractions *that* contained melanin were rejected   (3)

> The fractions, *which* contained melanin, were rejected   (4)

Sentence (3) tells the tale of a successful fractionation, (4) of an unsuccessful one. In (3), of n fractions taken (where $n > 1$), m melanin-containing ones were useless ($n > m$), but $n - m$ fractions were worth investigating further. In sentence (4) the *which*-clause states a property common to all fractions, and one that, alas, confers no "exclusive" tag. The implication is that we could have taken more fractions, but they would all have contained melanin. Observe – and this is crucial – that the *which* clause is separated off by commas. If these were missing, then the reader (who does not know whether the author is careful about his *that* and *which*) would be left dangling between hope and despair. Fowler puts it very well: "See commas safely through the press".

So far so good. In pairs such as (3) and (4), the distinction between *that* and *which* conveys much information. I shall virtuously try to observe it, and commend it to all my readers. On the other hand, I am not prepared to condemn anyone who confuses the two words as a despoiler of the precious heritage of our language. The distinction between the two words is of relatively recent origin, and is not part of popular speech in any English-speaking region.

So far we have only talked about the simplest case, the distinction between *that*-clauses where $n > m,$ and *which*-clauses where $n = m$. This distinction is so obvious that a careful speaker can make it instinctively. But matters get more complicated when $n = 1$. One should have thought that, where the main clause is in the singular ($n = 1$), the relative clause could not possibly be defining – what need is there to tie an "exclusive" tag to something of which there is only one? However, the logic of language decrees otherwise. Again I quote a pair of examples:

We were looking for a reagent *that* would remove the acetyl but spare the benzyl group   (5)

Paradigmanyl chloride, *which* fulfils this condition, is readily obtainable   (6)

Although sentence (5) does not explicitly say so, it nevertheless separates off from the set of reagents a subset of compounds having a specific reactivity. The *which*-clause in (6) performs no such function.

There are also sentences in which, at first glance, the reader is induced to believe that $n = m$. An example is

All *that* live must die   (7)

Here set and subset seem at first equal in number, but reflection reveals that, in a landscape filled with animate and inanimate things, sentence (7) places an "exclusive" tag on the former. Curiously enough, precisely in this type of sentence where analysis is difficult, speakers of English the world over never go wrong. We all say "all *that* remains" or "all *that* is necessary"; substituting *which* for *that* here is the kind of mistake *that* the English speaker is not capable of.

A good diagnostic test for the appropriateness of *that* can be made by inserting *those* in front of the subject of the main clause. If the sentence still makes sense, *that* is appropriate; if not, choose *which*. Thus, sentence (1) could be amplified by saying "*Those* compounds *that* contain nitrogen..." But no chemist jealous of his reputation could bring himself to say "*Those* alkaloids *that* contain nitrogen..."

One could be led to believe, from what has been said so far, that a rigid distinction between *that* and *which* was possible, and every editor should enforce it zealously. I shall now show that this is not so, and that a twilight zone exists between the two words. Let us write our amplified sentence (1) out in full:

> *Those* compounds *that* contain nitrogen play an important part
> in plant metabolism   (8)

Now *those,* in the English language, means only one thing. It informs the reader or listener that a subset is going to be split off from a set. Thus it usurps the function of *that* and the question arises whether *that,* which has suddenly lost its "defining" value, is still necessary or even permissible. Certainly no information is lost in sentence (8) if we replace *that* with *which.* I used to believe *those which* was fully as good as *those that,* and earlier versions of my articles abound with the former expression. Correspondence with W. E. Cohn has now convinced me that I was wrong. *That,* in sentence (8), has at the very least the merit of beneficial redundancy; its presence ensures that the information will come through even if the first word of (8) is garbled in transmission.

There are, however, occasions when the decision between *that* and *which* is so difficult that it cannot be reasonably expected a spontaneous speaker or writer will make it correctly. Let me frighten you with the following examples:

> In the type of sentence *that* is difficult to analyse, errors occur
> frequently   (9)

(*Those-that* sentences pose problems of logic.) In this type of
sentence, *which* is difficult to analyse, errors occur frequently   (10)

You see how hair-thin the distinction is? In (10), the "type of sentence"
has been so abundantly defined by the time you reach the relative pronoun
that there is no defining left to do; *that* must give way to *which*.

To sum up: the distinction between *that* and *which* is useful and can
generally (but not always) be made spontaneously. I shall try to develop a
finer ear for it and I hope you will too, dear reader. But if you slip up, do
not suffer agonies of guilt. As I have said before, the purpose of this book is
to help you enjoy your English, not to put you in fear of *which*-hunting
editors.

29 *A Package of Strings*

English is generously oversupplied with nouns of assemblage, such as "a *pride of lions*". Much snobbery and one-upmanship can be displayed in their correct choice; you lose status if you refer to an assembly of finches as a flock rather than a *charm,* or if you are unaware that a *gaggle* of geese on the water becomes a *skein* upon taking wing. It used to be a pastime of twittery humorists to suggest similar nouns for assemblies of persons. The English magazine *Punch* proposed a *flutter* of spinsters or a *fleece* of punters. The *Readers Digest* debated whether the personnel at Mme Tellier's should be known as a *flourish* of strumpets or an *anthology* of pros.

Let us leave this hee-haw of humorists to their own devices. Scientists are often accused of creating elitist in-jargon for the sole purpose of making outsiders feel ill at ease. If we were to level a similar accusation at hunters or naturalists, and to point out that a *hover* of trout and a *skulk* of foxes constitutes a waste of dictionary space, what an outcry there would be from lexicographers! These drab chemists, the world would be told, are trying to deprive the English language of its richness − let them leave our *wedges* of swans alone and stick to their conflagration of Bunsen-burners.

All right, lexicographers, the boot is now on the other foot. During the last century, the English language has been most wondrously enriched by words chosen by scientists to mean "a number of entities hanging together somehow". All these words are, more or less, synonyms of the standard English expressions *group* or *collection*. They are, however, words of more limited (and therefore more precise) significance, chosen not out of snobbery but out of necessity. Many of them, such as *system* and *ensemble,* come from thermodynamics; many come from statistics *(population, sample, configuration, complexion, array, distribution),* others from computer science *(package, matrix, string, file),* some from logic *(universe of discourse, set)* or from theoretical chemistry *(packet).* This profusion of terms cries out for a lexicographer who would codify the definitions, classify the terms, and thin out their ranks wherever he detected complete identity of meaning. The task is too ambitious for this humble essay. The definitions that follow are tentative, and subject to discussion and amendment by the scientific community.

A few household words first. Among the words that express "a number of entities hanging together somehow" *group* and *grouping* are perhaps the most general; theirs is the widest embrace. *Collection* conveys slight undertones of deliberate choice; we might say that we associate lower entropy with *collection* than with *group*. *Agglomerate* (and its synonym *conglomerate,* which is more fashionable at the moment) makes us think of heterogeneous members very tightly bound together; in an *aggregate* the tightly bound members are presumed to be similar. In a *cluster* (or its rare and highly forgettable synonym *congeries*) the members are not bound but close together in space.* *Assembly, assemblage* and *gathering* are close to *collection;* when used of persons they imply voluntary association.

Now come the specialist terms. *Set* is, to begin with, the scientific term for *group*. By choosing *set* we inform the reader that logical operations and a quantitative treatment are about to follow. Beyond that, *set* is the kind of group that can be generated by a logical process.

In thermodynamics, everything depends on precise definition. Authors of heterodox papers in this field are well advised to define even apparently simple terms right at the outset. I shall set the ball rolling by defining (for the purposes of this article only!) two terms of my own. I shall call THings the set of those things that can possess physical energy and matter; i. e. photons, subatomic particles, atoms and molecules. All other things I shall call thingS. Thus armed, let us analyse the word *system.* All standard texts use definitions containing such phrases as "a portion of matter being considered". I am worried by this, because *matter* excludes photons and *considered* drags in the observer. Moreover, I believe that *system* can have, according to context, two meanings; and I thus offer the following pair of definitions for discussion: "*system, n.* 1. All THings contained, at a specified moment, in a specified space. 2. The collective properties of these THings." Or am I wrong, and could one call a set of specified THings (say oxygen and methane) in a cylinder a *system,* while disregarding other THings (say nitrogen) also present? Over to you, thermodynamicists.

---

* This does not apply to organometallic chemistry where a *cluster* is a group of metal atoms bonded together.

There is no difficulty over *ensemble*. This is always taken to mean, in thermodynamics, the set of all possible states of a system.

Just as *set* warns the reader that the *group* being talked about will be analysed by the techniques of logic, the term *population* tips him off that he is about to enter the domain of statistics. A *sample* may be any part of a population that meets the statistician's requirement of random selection. An *array* is a collection arranged in some deliberate order. *Configuration* introduces the concept of geometrical arrangement; it also has the specialist sense of data being presented as *n* coordinates in an *n*-dimensional space. *Complexion* has a similar meaning: I do not feel qualified to define the word and anyway advise chemists not to use it. Mrs Murphy, during proof-reading, is bound to introduce a space after the *x*!

Universe of discourse is, in logic, exactly what *population* is in statistics. There is one faint difference between the terms: the former leads the reader to expect that the members of the group will be thingS, the latter hints of THings to come.

Computer people, in the first flush of their success, have created new terms with perhaps too gay an abandon. Logic is, of course, the computer man's second nature (he would say "first nature") and his new words are generally precisely defined and genuinely necessary. But, being an inverted snob, he has chosen to disdain jargon and instead adapts old household terms to new uses. This can create trouble when a word such as *package* falls into the hands of journalists. (Incidentally, we should not blame the computer boys too much. Ancient mathematicians have also re-adapted common words. Think only of *root*. *Matrix,* a term we now use for a rect-angular array, meant womb in Latin.)

Package, thus, is now a specialist term for "a group of entities which cannot be separated". A *string* refers to a linear arrangement of entities, so that you cannot by-pass B when going from A to C. (Incidentally, the physicists got hold of this term first.) A *tree* is another, and obvious, arrangement of a group, and so is a *network*. A *file* introduces the connota-tion of a group of like members being kept together for convenience; it may refer to data being made ready for input or to a grouping of programs to be used in conjunction.

Conjunction reminds me of such terms as *constellation* and *galaxy,* which are freely being used as synonyms of "group whose members are distinguished" in the popular press. We all wish now that these words had remained in the starry heavens. The men who used them first showed glittering imagination, but overuse has soon extinguished this radiance.

Thus, let us use our specialist words wisely. Let us keep their definitions precise, let us weed us unnecessary synonyms, and let us protest against their misuse by popularizers. Let us be a united band, a consensus of scientists, before we confront a frown of lexicographers.

30   A Clash of Symbols

My parents were born in a part of the former Austro-Hungarian empire, and as child I was much intrigued to see, on fading birth certificates and other documents, the prominent letters K. & K. It appears that in the old days the monarch was not only a *Kaiser* (emperor) of Austria but also a *König* (king) of Hungary, and both titles had to be duly recorded.

And thus I was exposed, for the first but not the last time in my life, to the spectacle of two letters K sitting side by side but meaning different things. In pre-war Austria, this apparently did not matter, indeed the fact that the two different titles had identical symbols may have been useful because it left touchy questions of precedence unresolved. Moreover, royal titles are not additive in the mathematical sense, and it would not have occurred to anyone to chisel "2K" over the portal of His Majesty's Water Tax Perception Office.

By now the reader knows what my coy opening is leading up to. And it is appropriate that we should have started with the letter K because it is the most abused symbol in the entire field of scientific communication. Suppose someone were to show you, out of context, the two characters 2K, and were to ask you to guess their meaning. How many guesses should you be allowed, to make certain of the right answer? Observe that we are talking only about *Capital roman* K, not about its italic or boldface brothers and its lowercase cousins.

Your first and best guess would have to be that 2K means two potassium atoms. Seeing your quizmaster shake his head, you might then venture the opinion that the 2K was in fact a badly spaced "2 K" and referred to a temperature of two kelvin. Unsuccessful again, you might then bethink yourself of a spectroscopic unit, surely the most unnecessary unit ever invented, called by coincidence *kaiser,* which is meant to be equal to 1 cm^{-1}. Wrong again? I can still think of one more possibility. The symbol K in computer operator's slang means 1000 or, according to the speaker's mood, 1024 ($= 2^{10}$) and chemists have begun to use this symbol frequently to indicate the extent of their data acquisition.

126

Klearly, this kalamitous akkumulation of K symbols must kome to a konklusion. But perhaps you disagree? You maintain that the context will make everything clear? My dear Sir, obviously you have not heard of Mrs Murphy's Second Editorial Law. Inexorable as only a Second Law can be, this proclaims: Anything that can be misunderstood will be misunderstood. (The First Editorial Law runs thus: All things that can go wrong will go wrong *at the same time.*) You know that the paper you are reading deals with radiation chemistry, and that the editor permits the use of kaiser. Thus you are certain that unit is meant when you see the sentence "Examplamine was irradiated at 1000 K". Very good. But Mrs Murphy will see to it that the irradiation is not carried out at 1000 K. If the numeral in front of K happened to be 293, would you still be sure? I have already mentioned (Chapter 25) a case where K for kelvin nearly insinuated itself into an equation describing properties of potassium. An alert editor will not allow the pages of his Journal to be sullied by such abominations as K for kaiser or K for 1024, but K for potassium as well as K for kelvin are IUPAC-ordained symbols, as are P for phosphorus and poise, V for vanadium and volt, S for sulfur and siemens.

The siemens is a term only recently coined by IUPAC to replace the mho, or reciprocal ohm, and IUPAC really should have known better. I implore all those august bodies that disburse research grants never, never to subsidize studies of the viscosity of yellow phosphorus, the potential changes on the surface of vanadium compounds, and the conductivity of sulfur.

Let me tell you another story. The Chemical Society once decreed, very sensibly, that all fundamental constants should be printed in boldface. But IUPAC has recently overthrown this by proclaiming that boldface was to be reserved for vectors and matrices. Thus the traditional k for the Boltzmann constant has to be printed in italic; the same symbol is also traditionally used for rate constants; and the two have a habit of popping up in the same equation. The *Australian Journal of Chemistry* gets out of this difficulty by insisting on the subscript B for the Boltzmann constant (k_B). But the world literature is full of equations in which k means one thing on the left-hand side and another on the right. Now you may say that this sort of thing would never trip you up. But it would trip up the computer into which the equation is fed.

Once I returned to the office in a state of replete somnolence, having just feasted a valued author at lunch, and found that Mrs Murphy had placed on my desk a manuscript in which the Boltzmanic produkt kT occurred in abundance. This was not underlined for italic and I rather mechanically marked in the underlining and the subscript B while trying to keep my wavering attention on the chemistry. Suddenly, I sat bolt (boltz?) upright. Among the many legitimate $k_B T$ terms, there was one that looked odd. Closer inspection revealed that no italics were called for; what was meant was a roman-type kT for kilotesla. It is moments like these that give an editor indigestion.

There used to be a *t*-honoured custom in all scientific literature that lower-case *t* always meant time, and *T* temperature. (Careful editors permit the use of *T* for *absolute* temperature only.) Heedless of this, the first investigators of nuclear magnetic relaxation *times* used T_1 and T_2 for quantities that could be expressed in seconds. They got these terms past somnolent (who am I to cast the first stone?) editors. At first this did not seem to matter, but pretty soon other workers got the idea of measuring the way these properties varied with temperature. Soon *T* for temperature rubbed shoulders with T_1 for time. I wrote to IUPAC and complained, but got the reply that T_1 and T_2 were "well established". So, of course, is the Mafia, but that is no reason for giving it official approval. May I remind the brilliant, courteous and helpful scientists on the appropriate IUPAC Committee that there is still *t* to put this mischief right?

I am not writing this article, dear reader, for the sole purpose of weeping on your shoulder. There are various things you can do to mute the clash of symbols. First, if your research is of such innovatory nature that new concepts are needed to express it, choose your symbols with care, discussing, where possible, your choice with the appropriate IUPAC Committee. Second, whenever you see a symbol that will cause confusion, complain to the editor and to IUPAC.

Remember that a symbol, to be acceptable, must fulfil various conditions. If it is a unit, it will be printed in roman type and may consist of several letters; it should not (as is "in." for inch) be identical with a word in common usage. If your symbol is a variable, it should have only one letter (italic or Greek) "on the line" but this may be followed by subscripts. Multi-letter terms such as VFA for percentage of volatile fatty acids or NOE for nuclear Oberhauser enhacement are not acceptable – to the uninitiated they seem not one symbol but the product of three.

Come to think of it, I am about to introduce a new unit myself. I have recently conducted some psychological research which allows me to quantify the degree of frustration felt by an editor when a symbol does not mean what it should. On my scale, 0 signifies absolute tranquillity and 100 the hurling of inkwells and ingestion of carpets. I had intended to call my unit the murphy, symbol M, but it occurred to me that M is already used in the literature for metal, molecular ion, and moles per litre. I have thus chosen the name kafka, symbol K. The name is in honour of Franz K., of Prague, 1883–1924, lawyer, insurance official, writer, and contemplator – half-humorous and half-anguished – of the innumerable unnecessary difficulties the human race so unfailingly creates for itself.

31   In Praise of Prepositions

Dictionaries grow ever fatter. The symptoms of lexical obesity are most noticeable in scholarly works. Where once a single hefty volume would dominate the shelf, there now stretches a trinity of tomes. In smaller dictionaries, whose sales depend on their handy size, the type size decreases stealthily from edition to edition, and the margins of the text grow closer and closer to the borders of the page. Is communication going to drown in a sea of words?

Well, no. Dictionaries are only approximate indicators of the growth of a language. A change in social customs, or an advance in technology, may create new words, but it will also render others obsolete. Yet these obsolete terms still remain in the dictionary; the worst that can happen to them is that an asterisk will mark them as archaic. Long after the last *farrier* will have hung up his hammer, the name of his trade will remain, a *furbelow* of language, a *fardel* to the dictionary that the publisher would *fain* get rid of. In the city of words, the dead are never buried, but remain in their houses – some even achieve resurrection, such as the beautiful chemical terms *moiety* and *handedness*.

But the fact still remains that the ocean of words is rising. It is rising because the icecaps of ignorance are melting. We have discovered more things, thought up more concepts, and explored the relations between these things and concepts more subtly than our ancestors. We need more words to communicate our discoveries and insights.

This is especially true of the Chemist's English. Our insights into the nature of the chemical bond, molecular geometry, and the magnetism of nuclei have pressed a vastly expanded vocabulary upon us. Many subsections of this dictionary still await their lexicographer, but I am glad to mention, in passing, that physical organic chemistry has found an excellent one in Professor V. Gold, who, under the auspices of the IUPAC Commission on Physical Organic Chemistry, has compiled a glossary of terms and published it in *Pure and Applied Chemistry* (Vol. 55, 1983, p. 1281). I warmly commend this document to all readers who may be puzzling, say, over the subtle distinctions between *stereoselectivity, stereospecificity,* and *regiospecificity.*

Now that we know something about the growth of language, we might ask how even that growth is. Does language grow (here is a word that would have puzzled Shakespeare) isotropically?

The obvious answer is that it does not. The greatest growth area is in the department of nouns. Verbs participate, to a moderate extent, in the noun boom, by virtue of the English peculiarity that every noun can be verbed and every verb nouned. Because every noun can generate an adjective, e. g., *stereoselective,* adjectives also contribute to the noun boom, and that drags along the adverbs. But how about the rest?

Well, all indicators show that the Word Exchange is bullish in the four word forms I have just mentioned, but bearish in all the rest. As far as pronouns go, this will cause no surprise: Thou knowest, O reader, that we have in fact lost one set of personal pronouns since Shakespeare's day. But you may not have realized that prepositions are also doing very poorly, and it is the purpose of this article to say that this is a bad thing.

31.1 The Function of Prepositions

Prepositions are mediators: They establish a connection either between a verb and a noun (he went *into* the laboratory) or a noun and a noun (the flask *on* the shelf).

In chapter 34 I shall propound a "chemist's grammar" in which the preposition P is described as the "electron pair" that creates a "covalent bond" between the verb V and the noun N; hence we have bonds such as $V(P)N$ or $N(P)N'$. Those subtle linguists who explore the translation of foreign-language texts by computers consider prepositions as "vectors" indicating the direction and intensity of the influence exerted by V on N, or N on N'. It might be better still to think of them as space–time coordinates that take you from an origin V or N towards the target noun. It is worth stressing that they are coordinates of time as well as space (the morning *after* the explosion, the building *on* campus) and sometimes one word can perform both functions (once he walks *through* that door, he is liable to work *through* the night).

The trouble with prepositions, especially English prepositions, is two-fold: There are too few of them and they perform well in explaining concrete relations but poorly for abstract ones. As to their numbers: I cannot think of a single full-blooded preposition that has entered the language since Shakespeare's day, yet some have been lost (for instance, *betwixt* and *athwart*). As to their performance: English prepositions acquit themselves very adequately when their task is just expressing relations of space and time. But when more subtle relations are involved – causality, emotion, logical dependence – it is hard to be precise without being cumbersome. How can we get the best possible value out of our prepositions? Let us take stock of what the dictionary has to offer.

31.2 Our Store of Prepositions

Let us call "preposition" everything that works like a preposition, i. e., everything that can perform the parenthesized function in *V(P)N* and *N(P)N'*. So defined, prepositions come in three shapes:

Thoroughbreds. These are the ones the dictionary labels primarily or exclusively as *prep.* – words like *in, over, after,* and *from.* Nearly all are short and have been with the language ever since it came to be called English. (A few latecomers, such as *re* and *via,* have forced their way into the dictionary and make occasional guest appearances in formal documents, but have no hope of ever becoming part of the spoken everyday language.)

Recycled participles. These are participles that, because of the shortage of genuine thoroughbreds, have been covertly reclassified as prepositions. Some of them *(pending, during, owing to, past)* have so completely detached themselves from the verbs that generated them that nobody thinks of them as participles; they are accepted as members of the Thoroughbred Club. Others (say, *concerning*) are close to such status; but with many, there is that microsecond of hesitation before the reader or listener recognizes their function in the sentence. Since thoroughbreds are adequate at expressing space–time relations, participles are mostly called upon to deal with more abstract matters *(regarding, notwithstanding).*

Frantic phraseology. These are word strings such as *with reference to, in respect of,* and *by means of.* Generally, the last item in such word strings is a thoroughbred preposition, and thus they often give the impression of a thoroughbred hitched to a heavy cart. Were it not for the shortage of unattached thoroughbreds, none of us would use these lumbering contraptions willingly. But unfortunately the supply of words in the other categories is limited, and sometimes we have to communicate *by means of* (try to replace that with something simpler!) such phrases.

Sorry to belabor the racetrack metaphor, but thoroughbreds are clearly the stylist's best bet if the right ones can be found. They are bred for the job — small enough not to overwhelm the words they are meant to join together, specialized enough in their grammatical function not to be mistaken for anything else. The trouble is that they are not horses for all kinds of courses: They do not perform very well on abstract tracks. For example, let, in two successive $V(P)N$ phrases, the meaning of (P) be first "as an undesired consequence of" and then "as a desired consequence of". We might come up with something like this:

> The experiment was postponed on *account of* an unexpected increase in pressure, but this was soon overcome *by means of* a leak detector. (1)

Can you think of any thoroughbreds to replace the italicized phrases? Yet the German language, even though its dictionary is slimmer than the English one, has two nifty thoroughbreds, *wegen* und *durch,* that can give exactly the performance required.

Most thoroughbred prepositions have acquired, as an extension of their space–time meaning, an abstract one. Whenever the transition from one meaning to the other is smooth (*beyond* the boundary, *beyond* endurance), such prepositions are a stylist's joy, because they endow an abstract concept with sudden visual impact.

There are, however, thoroughbreds whose abstract meaning has drifted away from the original one, and in such cases the visual impact is lost. Take *about*. In its original meaning, it was a synonym of *around* (the army lay *about* the city walls). In its present abstract use, it is nearly always vague and colorless: When we argue *about* a subject, we don't really talk our way around it but through it. Have you ever heard the hideous phrase *The discussion centered about . . .* ? That is like saying the center is located at the periphery. Replace such a horror with something like *The discussion had at its center . . .*

To sum up: Prefer thoroughbreds to all substitute prepositions. Always look for the most "pictorial" one. If you cannot find a simple preposition that suits your needs, don't agonize over your deficient language sense – the fault may be in your dictionary. Pass on, as we all must sooner or later, to the recycled participles or the phrases.

31.3   The Maintenance of Recycled Participles

As we have seen, certain words are, in the historical sense, participles. In the practical sense, though, they are prepositions, and we recognize them as such. Often the verb that gave rise to them has disappeared from the language (e. g., the parent of *during*, "to dure," has not endured) and thus there is no question of mistaking them for participles. Note how we can use the thoroughbred *in* for a passage of time if this time is intensely lived through (*in* my youth) but not when it is passively endured (*during* his illness, *pending* an investigation).

Besides the words that are unmistakably ex-participles, there are those that have a certain "tautomeric" quality: "I made a statement *concerning* Professor A" is most likely to mean that I said something about A to B; but it may possibly mean that I talked directly to A and caused him some concern.

Language contains a fair number of participle–preposition hybrids in which the prepositional tautomer predominates. Such words can generally be used without fear (for instance, if "Professor A" is replaced by "the safety valve"). But the price of clarity is eternal vigilance. Sometimes the unwanted tautomer will make a mocking appearance when least expected:

> *Regarding* nude bathers on the beach, the Reverend Ezra
> McBigot could scarcely restrain his emotion.   (2)

What can we say now *touching* those tautomers in which the prepositional form does not predominate, such as *considering?* Considering the ease with which they can be misunderstood, even greater care is required in their use.

Behind these "evenly distributed" tautomers now comes a vast crowd of "doubtful" or "spurious" tautomers. These are words that the dictionary describes as participles only, but that are being used with increasing frequency as prepositions. A good example is *assuming*. When we say "The yield was calculated assuming equilibrium conditions," we lay ourselves open to the charge of having broken the laws of grammar. But it is a sad yet true law of language that bad usage, if it is repeated often enough, becomes correct usage. Moreover, as we have seen, the slow transformation of a participle into a preposition is a common linguistic phenomenon. What's so special about *assuming* that it should not be allowed to share the fate of *considering*?

Having made these permissive noises, I would yet come out in support of the English schoolteacher, and beg you to rephrase your sentence as "We assumed equilibrium conditions for the calculation of the yield." The crime of the original sentence was not so much that it was wrong, but that it was fuzzy. Tautomers tend to give complex reaction products. At worst, we shall get such ridiculous by-products as Sentence 2. At best, we shall still have that undesirable moment of hesitation while the reader's mind searches for the noun the presumed participle is supposed to qualify.

All this is of great importance in the chemist's English because of the increasingly frequent occurrence, in the literature, of the "spurious" tautomer *using*. Compounds are recrystallized *using* chloroform; they are acylated *using* aluminium chloride; they have their diffractographic pictures taken *using* crystallography; their parameters are redetermined *using* the latest fashionable computer method. You name it, and some careless writer will put a *using* in front of it.

Most of the time the spurious preposition can be replaced by such thoroughbred prepositions as *by, with, in,* or *from*. At other times *using* replaces a phrase, generally *by means of*. Now it is unfortunate that the English language, generally so light on its feet, begins to plod when this particular concept has to be expressed. Thus the substitution of a simpler word for *by means of* may seem quite a good thing. The trouble is that *using* is being done to death by lazy thinkers. All would still be well if *using* were invariably used as a replacement for *by means of,* but unfortunately you find it used as a vague catch-all term meaning, more or less: "There is a connection between *V* and *N* but I don't know what it is."

My colleague Susan Ingham, who presides over the fortunes of the *Australian Journal of Zoology,* found the following sentence in an ecological paper dealing with Australia's most damaging animal pest:

Rabbits were observed using binoculars.   (3)

There, has that cured you? What a pity Walt Disney is no longer with us to draw the appropriate portrait of Mr B. "Binoculars" Bunny!

Expressions such as *by means of* and *with respect to* seem heavy-footed when compared to thoroughbreds and recycled participles. But sometimes they achieve a vividness of their own *(in accord with, with the aid of),* and at other times the paucity of our store of prepositions leaves us no choice. So use them unashamedly, but not too many in succession, please.

31.4 Concluding Remarks Anent Prepositions

It is still worth considering why the English language, perhaps the world's richest, should be so badly stocked in thoroughbred prepositions.

We users of English may have compounded our own miseries by neglecting such perfectly good words as *athwart.* What may have further contributed to the decline of the English preposition is the tendency of some modern newspaper editors to compress stories by squeezing all prepositions out of them. "Preposition Suppression Conspiracy is English Language Setback, Claims Australian Chemistry Journal Editor" – such a headline leaves a lot of strangled prepositions on the floor of the newsroom.

Finally, did you know that there exists, on the verge of extinction, a thoroughbred English preposition of impeccable history which nowadays has only an abstract meaning? If you are below a certain age group, chances are you will never have encountered the word at all. It is *anent.* It originally had the meaning "on an even (level) with" but in modern English it survived solely as a synonym of *concerning, with regard to, with respect to.* Fowler, in his "English Usage," tried his best to chase the word out of the living English vocabulary, claiming that it was in use in Scotland only, and everywhere else sounded affected. Some 20 years ago, *anent* might occasionally be found in querulous letters to the editor written by elderly correspondents; nowadays there is no trace of it. But it is a good word of honorable background, capable of strengthening the English language just where it is weakest. We chemists, unlike certain English lexicographers, are free of anti-Scottish bias. Could the word be rescued? Authors of good chemical papers who share my views anent *anent* will find the word encounters no obstruction in the *Australian Journal of Chemistry,* and I am sure my scholarly colleagues the world over will likewise be able to restrain their blue pencils.

32 A Prowl Among Personal Pronouns

Do you read French? Does your library stock *La Recherche?* Then do yourself a favor and read a delightful article by Jacques Pitrat (p. 876 of the October 1978 issue) called *La programmation informatique du langage.* Pitrat conducts research into artificial intelligence, and in particular into translations, by computer, from one language to another. It is doubtful whether computer translation will ever become a practical technique, but exploring its possibilities has led to some fascinating insights into language. You will find Pitrat's French easy to follow, except for one wildly untranslatable sentence: *Le professeur envoie le cancre chez le censeur.* Since we need this sentence as a starting point for our prowl among pronouns, we shall not wait for the construction of a computer capable of coping with it, but shall supply our own free translation to construct the following sentence fragment:

> The teacher sent Tommy to the headmaster because . . . (1)

Pitrat reminds us that this sentence could be completed in three different ways:

> . . . *he* could not tolerate his behavior any longer (1a)

> . . . *he* was throwing spitballs. (1b)

> . . . *he* wanted to have a stern talk with him. (1c)

Now comes the surprise. In Fragment 1a, *he* stands for the teacher; in Fragment 1b, *he* stands for Tommy; in Fragment 1c *he* stands for the headmaster. Yet all three completed sentences are correct! Eat your chips out, computer.

This raises the question of how the human mind can cope so effortlessly with a problem that would leave an ordinary computer program helpless. Let us sum up what we know about personal pronouns.

32.1   The Algebra of Language

Pronouns, as we have known since our own spitball-throwing days, are short words that replace nouns or noun phrases. Thus, they are merely convenient abbreviations; they do not add any new information to a sentence. The service they perform for us is similar to that which we obtain by introducing symbols such as *a or n* into our calculations. We may, for instance, define *a* as the expansion coefficient of a metal and then write *a T* to indicate its multiplication with the temperature. The two characters *a T* offer readers information that, were such symbolism unavailable, they could only gain from a lengthy string of words. Likewise, the pronoun "she" may serve as a symbol for a lengthy noun phrase ("the student I told you about − you know, the one with the glasses"), which otherwise would have to be tediously repeated.

Thus pronouns are useful things indeed, and even primitive languages have them. The step up from a pronounless language to one that has these particles is like the step up from arithmetic to algebra.

The trouble is that the algebra of language is bedeviled by a paucity of symbols. In the algebra of mathematics, we have symbols in abundance: we can choose among all the letters of the alphabet; we can capitalize them; we can print them in boldface; and if all else fails we can append subscripts to them ($a_1, a_2 \ldots$). In language the number of pronouns is restricted to begin with, and the laws of grammar allow us almost no choice among them. Apart from the choice between a few pairs such as *this/that* and *these/those,* the rules of language always dictate a unique solution for any given noun in any given situation.

Let us take English. This has three genders (the Romance languages have only two); hence, one has the three "algebraic symbols" *he, she,* and *it* available to replace any single person, thing, or concept that forms the subject of a sentence. In the case of the person being an object we have *him, her,* and − fatally again − *it.* But the choice between the members of these triples is not ours; it depends on the gender of the noun replaced.

Which is why computers don't like spitballs. What they would really like (they being the computers, not the spitballs) is an English language restructured as follows:

The teacher$_1$ sent Tommy$_2$ to the headmaster$_3$, because he$_1$
could not tolerate his$_2$ behavior any longer.   (1d)

Indeed, some computer translation programs use such subscripts or "markers" (about Pitrat's own more ambitious techniques, more in a moment). The amazing thing about human speakers in such difficult circumstances is that they (the speakers, not the circumstances) do not feel the need for such markers at all!

The textbooks on grammar that I have consulted remain strangely silent on the mechanism by which the human mind matches the pronoun with its referent noun. What happens, obviously, is that the mind retains a certain number of words in an "active area" of the memory and whenever a pronoun occurs, this area is searched for the appropriate noun.

The above is my amateurish language, not that of the professionals. Psychologists might talk of the retention of semantic units representing strings of words, and are wary of estimating how long such strings could be. Not only does the medium of communication matter, I have been assured, but also the length of time over which communication occurs. All one can assert is that the maximum number of intervening words is bound to vary from 10 to 100.

We can represent the search for the noun to occur in two phases: (i) a search according to "syntactic criteria" (i.e., the gender and number of the pronoun must match those of the noun), and (ii) a search according to "semantic criteria" (the match must make sense). (I do not know whether, in the actual psychological process, one search precedes the other; let us assume here that it does, in order to simplify our analysis.) Thus, in the case of the group of Sentences 1, the syntactic search establishes three possibilities. The semantic search then identifies for 1a the teacher as the only logical sender of Tommy; for 1b, the spotlight is thrown on Tommy as an investigator of spitball ballistics; for 1c, the headmaster is adjudged to have infinitely greater resources of sternness than Tommy. The pronoun *he* is matched to the noun accordingly.

Observe that, in view of the limited information available to us, a faint air of ambiguity still clings to Sentence 1c. For all we know, Tommy might be the crown prince of the all-powerful potentate of Petrolopolis, and he may want to visit on the poor headmaster his indignation at having found a dent in *his* (i.e. Tommy's) golden breakfast plate. Only the context can make clear who told whom off. So, let us talk a little about context.

32.2   Semantic Grammars

The grammar you learned at school is a syntactic grammar; it serves to test whether all components of a sentence are in their proper forms and proper places. Whether the resultant sentence makes sense or not is not its concern.

Linguists have often dreamed of creating a semantic or "sensical" grammar, a kind of grammar that declares absurd sentences to be illegal. The computer programs described by Pitrat are a step in this direction. These involve attaching semantic "markers" to nouns and verbs (thus a noun might be classified as "animate" or "inanimate" or "abstract") and developing "background scripts" for various areas of activities (school-room; teacher instructs; has authority over pupils). Whenever these markers do not match (say an abstract noun becames the object of the verb "to eat") or an activity departs from the background script (classroom is venue for heart transplant surgery) the sentence being tested is signaled as doubtful.

Let me mention, in passing, that there exist languages whose grammars have semantic features. In Chinese, for instance, every noun is preceded by a "classifier," which assigns *it* (the noun, not the classifier) to a certain category (persons/animals/plants/flat inanimate objects/bales/piles/ bundles, etc.). These classifiers, no doubt, became part of the language for phonetic reasons. There are many nouns in Chinese that are homonyms, and the classifiers distinguish them. But they also act as semantic markers in Pitrat's sense; for instance, a sentence in which an inanimate object governs the verb "to think" will be rejected as ungrammatical. It seems Confucius, whom we were so sternly invited to criticize during the Cultural Revolution, knew a thing or two after all.

An interesting feature of semantic grammars is that they do not test absolute correctness of sentences, but correctness relative to the intended audience. The hallmark of success, for a semantic grammar, would be its ability to spot not only absurd but also ambiguous sentences. Now take the group of complete Sentences 1. A semantic grammar will classify them as correct only if they are addressed to an audience of a certain cultural ambience. If our listeners or readers have no means of knowing that teachers have authority over pupils, and that the latter are more apt to throw spitballs than the former, then the sentences are unacceptable. It is an uncomfortable fact of life that what is good English in Peoria may be bad English in Petrolopolis!

32.3   The Chemical Context

So far, we have conducted our investigation in terms of everyday English, but the application to the Chemist's English is obvious. To know whether our sentences are "correct" or not, we have to make a precise assessment of the background knowledge possessed by our prospective audience. Let us create an analog to Sentence 1:

> The acid damaged the lining of the reactor, and *it* had to be re-placed. (2)

The teacher, his attention no longer distracted by Tommy, will confirm that Sentence 2 meets the requirements of syntactic grammar. But how about semantic grammar? Let us write the sentence with our own markers in place:

> The acid$_1$ damaged the lining$_2$ of the reactor$_3$, and it$_n$ had to be replaced. (2a)

Can we assign a definite meaning to n, the marker of *it*? As long as we do not know the context, we cannot. All we can say at the moment is that there is a strong and equal likelihood of n being 2 or 3, and a very small one that it is 1. (It may be that the acid is simply required as a catalyst. Having used sulfuric acid and discovered that it chewed up the lining, we may have quite successfully replaced it with formic acid.)

Now all this may change if we know the context. The reactor may be a huge and immobile thing, or else a tiny cell that can be plugged into or out of a gas stream; knowledge of this might lead to an almost unambiguous assignment of n as either 2 or 3. But is "almost" good enough, and anyway how clued-up is the author's audience about chemical engineering? *Any sentence that can conceivably be misunderstood by its intended audience should be rephrased.* For instance, if n is 2, we might be well advised to write:

> The acid damaged the reactor lining, and it had to be replaced. (2b)

This highlights the link between *lining* and *it* because *reactor* has now become a noun-acting-as-adjective and as such is not really entitled to be symbolized by a pronoun. (It seems to me a law of syntactic grammar that adjectival nouns cannot be linked to pronouns. At least, I have been unable to construct an acceptable sentence in which such a linkage occurs. Could you live with "The moonlight shone on the palace portal, and its windows glistened"?*)

32.4   The Double-Edged "It"

Nearly all the nouns in the chemist's vocabulary are of neutral gender, and that introduces a further difficulty. It so happens that the accusative form of the neutral singular pronoun *it* is identical with the nominative. For readers who protest at this technical gobbledegook, I shall explain by giving an example:

> The nitrate$_1$ did not, in fact, enter the aromatic structure$_2$;
> *it*$_1$ oxidized *it*$_2$.   (3)

Two *its*; one a subject and one an object; and yet we can confidently assign each to its appropriate noun. Our confidence is sustained by the background script, which tells us which species is the oxidant. For the proper matching of nouns and pronouns, we also depend on the fact that the sequence 1, 2 is the same in both clauses. Let me prove this point by an example from everyday English.

Suppose that you are writing a historical treatise about the unsuccessful attempts of a small neutral country to coexist with a large bellicose neighbor. You might very well end a chapter with:

> In the event, the lion$_1$ did not lie down with the lamb$_2$; it$_1$ ate it$_2$.   (4)

A perfectly clear sentence, and an effective one. Yet try to invert the 1, 2 sequence by substituting "... the lamb$_2$ could not lie down with the lion$_1$ ..."! The sentence collapses, even though the background script tells us loud and clear which beast is the carnivore.

---

* My colleague D. E. Boyd promptly put the skids under this theory of mine by pointing out that it certainly did not hold for demonstrative pronouns. He constructed the sentence: "Safety authorities recommend yellow$_1$ raincoats for children because extensive tests have shown that this$_1$ is the most visible color under all climatic conditions." I leave the reader to ponder whether this sentence would be acceptable if the demonstrative pronoun *this* were to be replaced by the personal pronoun *it*.

It is worth noting that this difficulty is only caused by the double-edged *it*. Wherever a pronoun can be uniquely assigned to a noun, the 1, 2 sequence can be reversed:

Michael's$_1$ reputation$_2$ did not put Carla$_3$ off; she$_3$ disregarded it$_2$ and married him$_1$.   (5)

We can only hope that they lived happily ever after, untroubled by pronouns.

33 How Good is Your English?
How Good is English?

Has any language ever dominated the world of science so much? The number of chemical journals that do not accept contributions in English is shrinking rapidly. And once an editorial committee decides to allow the use of English in the pages of its journal, if finds that it has invited a cuckoo into its nest that pushes the native fledglings aside. These days, a knowledge of English will get you safely through *Acta Chemica Scandinavica* or *Gazzetta Chimica Italiana*. A look at *Helvetica Chimica Acta* will make you think that there are entire Swiss cantons where English resounds from every mountaintop. Moreover, most of those major journals that are still published in a non-English language either offer their own translations to the international scientific community *(Angew. Chem., Coll. Czech. Chem. Commun.)* or have that job done for them by outsiders *(Doklady)*.

Did this marriage between chemistry and English take place because the bride was so beautiful, or because her parents were so rich? It must be admitted that the wealth of the parents had something to do with it. Consider some of the bridegroom's past flirtations, and how they were discreetly abandoned when family fortunes declined. The present bride certainly brings with her a resplendent dowry: the huge American research budget, the substantial contributions from Great Britain, Ireland, Canada, and Australia. But it would be most ungallant to deny her personal charms and virtues.

At this stage, let us abandon the metaphor of the beautiful bride. Beauty contests between languages are just as invidious and foolish exercises as the other kind. What matters, and what explains the present dominance of English in the chemical literature, is that it is a highly efficient communications system.

We all know the three virtues a communications system must possess. It must be easy to encode, easy to transmit, and easy to decode. If we draw up a report card for English, we find it gets good marks in each subject. Good marks, but not full marks; it is the purpose of this article to draw attention to certain imperfections of the English language that we must consider in encoding, lest our message be misunderstood by the decoder.

33.1   Transmission

Before we lose ourselves in subtleties of grammar, let us test how easy it is to transmit. Here it is at the head of the class. Its alphabet has only 26 characters. (Some publishers, it is true, combine the first two letters of *ae*sthetics into a special character, and others place a diaeresis − the double dot − on the second vowel of coopt, but such practices are uncommon.) Nearly all other languages need either larger keyboards or else more keystrokes for transmission of text. French, for instance, possesses besides the unaccented letter *e* the three additional characters *é, è,* and *ê*. The Russian alphabet has 32 letters (33 if the two tonal values of the letter *e* are distinguished by a diaeresis). And as for the other main contender of English, Japanese, I beg to be excused from making a character count.

33.2   Encoding

It can readily be shown that written English is an easy language to encode; however, let there be no premature rejoicing over this. The energy saved in framing a simple message is often expended again, with interest, at the decoding end. Compare, for instance, binary and decadic arithmetic. The former is much "easier to learn", and of course it is easier to transmit. But who of us would like to puzzle out a gas bill for $ 110110?

Made cautious by such thoughts, let us now explore the advantages of English as an encoding system. Its grammar has just the right kind of simplicity. True, there exist major languages with even simpler grammars, e. g., Chinese. But in such languages it becomes difficult to construct sentences of the complexity, for which the sentence now in progress affords an example, required for scientific communication.

So, let us count our blessings. In English, if you know the meaning of a noun, you know, with 100% certainty, its gender. (In German, you have to work your way through a set of complex rules before you can aspire to something like 75%. How are you to guess that *table* is a he; *door,* a she; and *house,* an it?) If you know the singular form of a noun, you know, with 99.9% certainty, its plural. Can you tell me, offhand, the plural of that well-known German immigrant into the chemist's English, *Festschrift?*

There are, perhaps, 200 irregular verbs in English. To manipulate the others, all that is required is a knowledge of the endings *-s* and *-ed,* plus some auxiliary verbs. French and Russian grammar books bulge with tabulations of complex verb forms. In the Semitic languages, verb endings change according to the gender of the noun. Latin languages have an additional "literary" past tense, which is used in writing but practically never in speech.

English adjectives never change their form. In most other European languages, their form depends not only on the gender of the noun but also on its number. Had enough? Or shall I terrify you further by explaining the distinction between dative and accusative that is made by all other European languages but is practically lost in English?

It follows that it should be much easier for, say, an Indonesian to learn English than to learn Russian. It will also be much easier for a Russian to learn English than for an American to learn Russian. Not only will the American have to learn an unaccustomedly large number of word forms; he or she will also have to learn a novel set of grammatical concepts that generate the rules controlling the word forms. In the Tower of Babel, as in any other tower, it is easier to walk down the stairs than up; and what contributes to making the English language so readily accessible is that it occupies one of the bottom floors.

33.3   Decoding

The price of simplicity, however, could easily be obscurity or ambiguity. Is English harder to decode than any of the more complex languages? My own reassuring answer is: only marginally so. We pay a very small price for a very great gain. Let's look at what "decoding" involves. The reader first identifies the noun phrase (and in it the subject noun or pronoun) and the verb phrase. In the verb phrase, the reader notes the state of the verb and its direct and indirect objects (if any). The reader looks out for subordinate sentences and assigns them to the noun phrase or verb phrase as the case may be, then identifies the referents of all pronouns, and establishes the proper connection between words closely bonded by meaning but separated by strings of words in the word sequence (say a verb and its adverb; a noun and a participle).

The whole operation sounds like a fiendish classroom drill under the direction of a cane-wielding schoolmaster. Instead, it has become as natural to you and me as walking. And even as you decode this sentence, countless Chinese readers perform an essentially identical feat in front of their ideographs, and countless Russians sail with equal ease through texts printed in Cyrillic script.

Are we English readers at a disadvantage compared to such confreres? I know of no genuinely multilingual person (and I have met many) who ever said reading in English slowed him or her down; in fact, some have affirmed the contrary. Nevertheless, the great simplicity of English sentence construction does occasionally lead to ambiguous phrases: The encoder thinks the meaning is beyond dispute, and fails to realize that another meaning is also possible. To forearm ourselves against confusion, let us list some minor glitches in the English communications system.

33.4   Homographs

Homographs (two words of identical spelling but different meaning, and often of different pronunciation) bedevil other languages, too. (For instance, in Italian *ancora* may mean "again" or "anchor"; *subito,* "quickly" or "undergone"; *pesca,* "fishery" or "peach".) English has more than its fair share. Think only of such sentences as *The actress's entrance will entrance,* or *A number of my fingers grew number.* In the chemist's English, the discourse will *lead* to *lead.* One really annoying blemish of English is the identical spelling of the present and past tense of the verb *to read.* Occasionally, to protect the reader, it may be advisable to emphasize the past tense by an adverb such as *recently.*

33.5   Formal Equivalence of Verb and Noun

Unless a noun has a telltale ending such as *-ation,* or a verb ending such as *-ize,* there is no certain way to tell such words apart. In fact, many nouns do extra duty as verbs, and vice versa. The practice of verb-nouning and noun-verbing goes back to post-Chaucerian days, when the ending denoting the infinitive (*work-en, drinc-an*) disappeared from the language. The Elizabethans, especially, broke down the barriers between noun and verb, with often very exciting results. These days, however, the mix of nouns and verbs has gone as far as it should be allowed to go. Occasionally the reader is slowed down by the difficulty in finding the predicate verb (verb and noun mix makes decoding difficult), and it is possible to encounter genuinely ambiguous headlines such as

Mayor suggests housing development works.   (1)

Either *works* is a noun and in that case His Honor is proposing that new developments be undertaken; or it is a verb, in which case he is congratulating himself on the smooth functioning of an existing undertaking. The equivalent in the chemist's English might be

X-ray tests show crystal forms.   (2)

The moral of all this is that the chemist should not add too readily to the existing store of verb-nouns. Instead of saying *cobalt complexes readily* we should say *cobalt forms complexes readily.* I am not suggesting, as some purists do, that we should abandon the verb *to reflux;* it has become part of our language. But don't let this thing snowball. If a noun has to be coined from a verb, use the gerund (see below) or use a suffix such as *-ion* or *-ation.* For verbs from nouns, try *-ate* or *-ize.*

33.6   Formal Equivalence of Gerund and Present Participle

Forming a gerund from a verb is a perfectly legitimate way to turn the verb into a noun. I delight in letting my fingers run over a typewriter, hence

Writing is my favorite pastime.   (3)

Here, the gerund *writing* is the subject of the sentence; in fact, it is a one-word noun phrase.

Rather unfortunately, the English gerund is identical in form to the present participle. Generally, this causes no trouble; the gerund nearly always announces its noun status by being preceded by a preposition or article (whereas a participle could never be found in such company). *After reading* this explanation, the reader will have no difficulty with *the unraveling* of this sentence. However, ambiguities can occur. Chomsky has given a well-known example:

> Flying airplanes can be dangerous.   (4)

If *flying* is a gerund, then the pilot is in danger; if it is a participle, then the threat is, perhaps, to a bird about to be sucked into the jet intake. I must admit, though, that the danger resulting from this particular weakness in English grammar is more for the birds than for chemists. In more than two decades of editorial activity I have never encountered a genuinely confusing sentence of this type. Anyhow, you have been warned. *Disdaining* advice can cause a lot of trouble.

33.7   Formal Equivalence of Past Tense and Past Participle

The ambiguity that follows was brought to my attention by Professor N. V. Riggs of the University of New England. (American readers will be interested to learn that Australia, too, has a New England region; it lies about halfway between Sydney and Brisbane.) Consider the following sentence:

> The aqueous layer was removed, and the solvent concentrated.   (5)

No ambiguity here. To avoid tedious repetition of the auxiliary verb *was,* it has been omitted – quite legitimately, according to usage and all the textbooks – in front of *concentrated.* Now let us expand the sentence:

> The aqueous layer was removed, the solvent concentrated, and
> the residue exploded.   (6)

Here, as we remove our safety goggles with shaking fingers, we find ourselves in trouble. A grammarian, insisting that consistency in sentence construction must prevail, will interpret the sentence in one way only. The auxiliary *was* having been omitted in front of the participle *concentrated,* we are compelled to conclude that it has been omitted in front of the participle *exploded* also. The meaning is thus: . . . was concentrated . . . was exploded. But *was exploded* denotes a deliberate act. What we had meant to say was that *the residue exploded* out of its own inscrutable malice, not by our deliberate intent. How are we going to make this clear to the Dean, who even now glares balefully at the shambles in the laboratory?

My advice is this: Whenever this confusion threatens, don't flirt with disaster but put in the auxiliary verbs. We don't want a semantic shambles on the writing desk any more than broken glass in the fume cupboard.

33.8   *What Else is There?*

In Chapters 11 and 31, I wrote about the confusion that can occur when a preposition having the form of a participle suddenly reverts to its participial function. I gave some funny (I hope) but also rather contrived examples.

Have I forgotten any other source of ambiguity? Is there another occasion on which I should have caught the English language napping? I cannot think of any. It seems to me the English language has passed its examination rather well. It is easy to encode, easy to transmit, and there are no major problems in the decoding process — provided, of course, that the encoder takes a little bit of trouble over his message.

34 A Chemical Analysis of the English Sentence

Page 15 of my copy of the "Handbook for Chemical Society Authors" contains a paragraph of three sentences. I shall number them here for convenience:

1) "React" is an intransitive verb only.
2) "X reacts with Y" is correct; "X was reacted with Y" is incorrect and must be replaced by "X was treated with Y" or "X was caused to react with Y".
3) Similarly, "unchanged acid" cannot be rendered "unreacted acid".

These three sentences draw attention to a very common, but not very serious, error in scientific writing; in a good many journals the mistaken use of *to react* goes unnoticed or at least unchallenged. But it is a challenge to an alert copy-editor and of course it is dead easy to correct − two strokes of your pen, and "reacted" has become "treated". Thus my colleagues and I have always taken pride in righting this wrong done to the English language.

But what if the author does not agree he was in the wrong? Or if he wants an explanation why a phrase that he has come to consider as natural is not good English? It is easy to assert that a certain usage is wrong, but it is hard, very hard, to prove it. The misuse of the verb *to react* is a trivial matter but, in trying to explain that it is a misuse, I found out a good many non-trivial facts about language. In what follows I shall try to construct a "chemist's grammar", that is, a method of analysing sentences that uses the concepts, and the deductive techniques, of the chemist. It is just a primitive little grammar, not made for the analysis of highly complex sentences, nor is the idea (as I found out halfway through) original, but I had much fun with it and often find it useful.

My immediate objective will be to prove something that many of my author-friends disputed: (1), (2) and (3) are not three separate instructions, but once you accept (1) then (2) and (3) follow logically. First of all, though, I have to ask you, dear reader, whether statement (1) has any meaning for you at all. Chances are that if you are under thirty and have received your education in an English-speaking country, you will not even know what an "intransitive verb" is. If so, please don't give up on this article; all that needs to be explained will be explained, in terms familiar to the chemist.

All that needs to be explained, that is, except for two concepts from traditional grammar that everyone knows well: the terms *noun* and *verb.* These two terms retain, in our chemist's grammar, their familiar meanings of an-object-or-concept and the-kind-of-action-performed, respectively. We shall also import from traditional grammar the term *preposition,* and shall define it in terms of our new grammar as follows: A preposition is a word or phrase that creates a link between a verb and a noun (I looked *into* the microscope) or a noun and another noun (the bottle *on* the shelf). Apart from *verb, noun* and *preposition* we shall start from scratch.

I have already sketched out the principle of the new grammar in the Preface. We shall consider the English sentence as a molecule and its words as the atoms, and then we shall investigate the nature of the bonds that link these atoms.

What kind of molecule had we best compare the English sentence to? Obviously one that contains a great number of bond types. My favorite molecule is that of a coordination complex, because it has a dominant atomic species, the metal, that offers a perfect analogy to the verb in the sentence. Let us imagine such a complex, and write down down an idealized reprentation of it as shown in Figure 1.

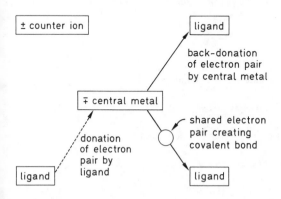

Fig. 1. A multitude of bonds.

The Figure reminds us that there are various ways to accomplish the traffic in electrons that we call a chemical bond. There is, first of all, the ionic bond, which might be described in layman's language as the permanent loan one atom may offer to another. Then there is the covalent bond, which is, as it were, a partnership of two atoms investing equal shares to create an electron pair as working capital. Next comes the coordinate bond, with a ligand atom "providing" an electron pair to the metal; this bond is shown as a broken line in the Figure, to distinguish it from the one to be discussed next. This is the back-donation bond, in which it is the metal acting as the "banker" that lends an electron pair to another ligand.

One central component, linked to several peripheral components by a variety of bonds – what is true of our idealized molecule is also true of the simple English sentence. Let us draw our picture again, with one change of symbol and introduction of a new one (the wavy arrow, see below): I shall replace the plus-or-minus signs with male and female signs because the analogy between the ionic bond and the (subject noun)–verb bond is not as close as the other analogies. Figure 2 gives us our first view of the "sentence-molecule" and the names we shall use for the bonds.

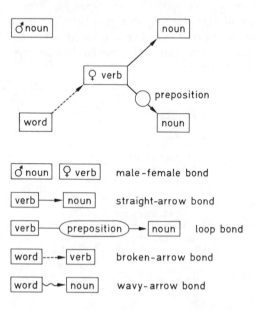

Fig. 2. The tools of the chemist's grammar.

Note the simple names given to the various bonds, and note also a newcomer, the wavy-arrow bond. This differs from the others in that it is never linked to the verb but can be linked to any noun in the sentence. Note also that the name *noun,* in our primitive grammar, also includes pronouns. Let us now investigate each type of bond in turn.

The *male-female bond* links that-which-performs-the-action (the "subject" of conventional grammar) to the verb. As the name indicates, here the meaning flows from noun to verb; we want to know who or what acts before we can comprehend the action. Thus ♂ Bob ♀ wrote . . .

In the *straight-arrow bond,* the meaning flows from the verb to that-on-which-the-action-is-performed or what-the-action-achieves: wrote → a sentence . . .

In the loop bond, the preposition, pictured as sitting in the loop, imitates, in an amusing way, the part played by the electron pair in the covalent bond: it draws two nuclei of meaning together, with the sense flowing away from the verb: wrote —(on)→ the blackboard . . .

In the *broken-arrow bond,* the meaning flows towards the verb, as in slowly ----► wrote . . .

In the *wavy arrow,* the meaning seems to flow towards the noun: boring ～►sentence, dusty ～► blackboard

Figure 3 shows a complete simple English sentence. Note the letters *a, b, c, d* placed next to the four principal types of bond. We shall investigate how many bonds of each species the English sentence can contain.

Let us first adopt the convention that, if after any word in our sentence we introduce the word *and* followed by another word, the number of bonds

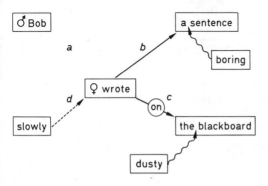

Fig. 3. Bob's activity revealed!

is not changed thereby. For instance, if we replace "Bob" by "Bob and Joe" or "wrote" by "wrote and explained", we do not create a new atom in our molecule but, so to speak, replace an atom by its isotope. That settled, we can now address ourselves to the task of finding the number of male-female bonds *a*.

No problem here. Pick any simple sentence in the Chemist's English at random, and you will invariably find that *a* equals 1. I specify that I am talking about the Chemist's English, by which I mean the sort of language you would use in the Discussion section of your paper. Grammarians classify exclamations such as "Careful!" or "Oh yeah?" as sentences, and some of the text of your Experimental section might be expressed in telegraphic style, or contain an imperative (*heat slowly*). But here we are not interested in such imperfect sentences. We can affirm that every sentence in the Chemist's English must contain a verb, and *one* "male" noun, and if this condition is not met we do not have a sentence.

Now for the number of straight-arrow bonds *b*. We find that this must be either zero or one, and that no other value is possible. At first you will not be ready to accept that statement, because there are many English sentences for which we find, apparently, *b* = 2, such as *Fred gave Bill a book*. But this is a delusion brought about by what scholars call the disappearance of the English dative. It is a property of the English language that if you do something *to* someone, such as give, offer or tell, the *to* can sometimes be omitted. So (Figure 4) the supposed second straight-arrow bond stands

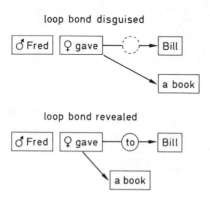

Fig. 4. The vanishing English dative.

revealed as a disguised loop bond, with the phantom preposition *to* sitting in a phantom loop. That this preposition is actually intended can be revealed by changing the order of the words: *Fred gave a book* to *Bill.*

Now that this apparent anomaly is explained, we can affirm that there are two, and only two, kinds of English sentence possible, those with $b = 0$ and those with $b = 1$. The $b = 1$ state is, as it were, an additional valency the verb may or may not acquire.

This yields us the first practical application of our exercise, because it provides us with a handy definition of a difficult term in conventional grammar. Verbs in $b = 0$ sentences are called *intransitive,* those in $b = 1$ sentences *transitive.* We shall appropriate these terms for our grammar and make good use of them, but first let us complete our quantitative analysis of the other types of bonds.

We find that the number of loop bonds in a sentence is theoretically unlimited. Bob wrote a sentence *on* the blackboard, *at* night, *with* great effort, *in* a bold hand ... Likewise there is no theoretical limitation on broken-arrow bonds. Bob might write *slowly, carefully, agonizingly, ...,* all in the same sentence. But in practice there is a limit on the numbers c and d. Too many of the bonds they represent make "the molecule unstable". In the chemist's English, if you go beyond $c = 4$, $d = 3$ or $c + d = 5$ your sentence is likely to be too ponderous. (Of course a great stylist may defy this rule of thumb and get away with it.)

But let us return to the study of the straight-arrow bond. Go to the dictionary and start picking verbs at random. You will find that nearly every English verb has a definition for *v. t. (verb transitive;* our $b = 1$) as well as one for *v. i.* (our $b = 0$). A huge majority of English verbs can acquire the additional valency represented by the straight-arrow bond and readily lose it again. Look only at the sentence depicted in Figure 3. If we reduce its state to $b = 0$ by eliminating the straight arrow and all the words dangling from it, the sentence remains perfectly acceptable: *Bob wrote slowly on the dusty blackboard.*

It is very hard indeed to think of verbs that do not display both valency states. I cannot think of a single verb that is exclusively transitive, though for practical purposes words such as *use* are hardly ever found in the $b = 0$ state. One always uses *something.* There are, on the other hand, just a few verbs that are exclusively intransitive, such as *to arrive* or *to result.* You cannot arrive somebody or result something. Among these few verbs is the one that interests us, *to react.* In everyday language we instinctively know this to be true. We can say: *The management reacted to the strike threat.*

But we definitely cannot say: *The management reacted the strikers with a lock-out.* None of us could possibly make such an error in everyday English.

The user of the Chemist's English, having followed the argument so far, is likely to accept the evidence of the entry "react, *v. i.*" in the dictionary, and to concede that he should not use the phrase *We reacted examplanoic acid with diazomethane.* But that is not a phrase he uses much anyway. He writes, in the passive voice, *examplanoic acid was reacted with diazomethane,* and he feels perfectly comfortable with this expression.

I am sorry to deprive him of that comfort: the expression violates a very basic law of language. Let us investigate the way the active English sentence changes into the passive. First of all, by ways our primitive little grammar is at this point unable to codify, the form of the verb undergoes a change; for instance *wrote* becomes *was written.* Then, just as a chemical molecule might contort itself at the approach of a reagent, our sentence-molecule "flips about" and its bond structure undergoes the changes shown in Figure 5.

Consider first the sentence-molecule with a transitive verb, i. e. the one that possesses a straight-arrow bond. The noun at the end of this bond becomes the male component of the male-female bond in the passive sentence. The former male-female bond disappears and is replaced by a loop bond in which the preposition *by,* or one of its synonyms, becomes the "electron pair". All other bonds remain unchanged.

But what if the verb is intransitive? The molecule then does not contain a straight-arrow bond. It follows there is no noun available to act as male component; thus a will be zero when it should be one; and we have already seen that in that case we have no sentence. In other words there exists a law of English grammar, in fact of universal grammar, that states: *A sentence containing an intransitive verb cannot have a passive form.* (I must confess that I was totally unaware of this law until I started tinkering with my chemist's grammar.)

We have thus proved that proposition (2) at the beginning of this chapter can be deduced from proposition (1). "React" is an intransitive verb, and *therefore* "X was reacted with Y" is incorrect. It remains to prove – and here I have encountered the stiffest opposition from authors – that (3) follows also.

Let us apply the technique we have used to analyse the active/passive transformation to the verb/participle change also. Participles, as we have seen in earlier chapters, are always attached to nouns; in terms of our

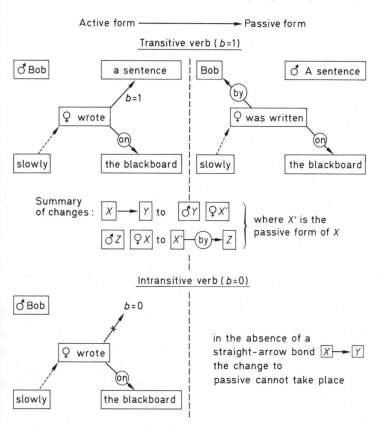

Fig. 5. The "flip" from active to passive. *X, Y, Z* are abbreviations for words or groups of words.

grammar that means that they always stand at the beginning of a wavy-arrow bond. Let us recall that, to form the present participle, the verb stem adds the ending *-ing,* and for the past participle the ending is mostly *-ed* but sometimes *-en* or *-t.* The verb/participle flip (Figure 6) leads to two different bonding patterns according to whether the end product is the present or the past participle.

The present participle, *X*-ing, always forms a wavy-arrow bond with the former male component; in terms of Figure 6 we could say that the present

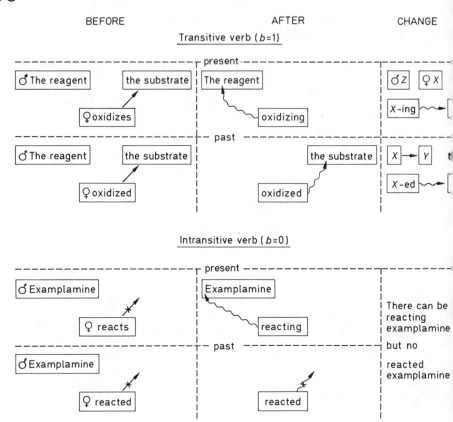

Fig. 6. The "flip" from verb to participle.

participle always "bonds to the left". The past participle bonds the other way, "to the right"; where formerly there was the straight arrow a wavy arrow now appears. Thus we have oxidi*zing* reagent but oxidi*zed* substrate. But in the sentence "examplamine reacts" there is no straight arrow that could be changed into a wavy one. Thus the past participle of an intransitive verb cannot attach itself to any noun, and if you cannot have a "reacted substance" then obviously you cannot have an "unreacted" one. Take out your pens, scan your last manuscript, and transform any *unreacted* into *unchanged*. *To change* is a perfectly normal and uncomplicated English *v. t. & i.* verb: your attitudes may *change* ($b = 0$) and as a consequence you may *change* your habits ($b = 1$).

One could not call what we have just demonstrated a law of universal grammar, because there are many languages without participles, but in all languages that have these word forms the law is valid. In English there is the additional nuisance that the past tense and the past participle are generally formally identical; but if you analyse a sentence by means of the chemist's grammar the presence of the wavy arrow will always reveal the participle.

Over the years, I had often toyed with this "my" grammar, and finally decided to write my system up for a lecture tour, hoping my listeners would be amused by all the blackboard tricks. But as I committed my ideas to paper, I developed an uneasy feeling that my ideas were not new. It occurred to me that I had constructed a *transformational grammar,* a very humble cousin of the elaborate systems of high intellectual distinction that Noam Chomsky has been developing over the last three decades. The purpose of such grammars is to chart the processes by which a simple affirmative sentence is changed into an interrogation, an imperative, the statement of a possibility and so on. The grammar we have developed in this chapter is far less ambitious, but it is readily accessible to chemists and others educated in the physical sciences, and it has its limited practical uses – as we have seen.

I realize, of course, that I have brought far too elaborate machinery into play to deal with the venial sin of misusing *to react*. If, in the end, the use of *to react* as a transitive word becomes too common, the dictionaries will give in and list it as such. In language, established usage comes first and logic second. Think only of what happened to the verb *to distill*. Its literal meaning is "to come down in drops". Thus it is logically correct to say "the ether distilled" but not "we distilled the ether". Yet usage has won this battle and both forms are equally acceptable. In other words, I am ready to allow you to commit, for the verb *to distill,* the very same error of logic that I have just so elaborately forbidden you for the verb *to react*.

But I make no apologies for that. *To reflux* and *to distill* are words almost exclusively used by chemists; what we do with them is our own affair. But *to react* is in general use, and if we corrupt logic here the corruption will spread.

And of course the use of the chemist's grammar is not restricted to this trivial case alone. First of all, it is a lovely teaching instrument for sentence construction. And there are other practical uses. For instance, let us enlarge our grammar so that it extends to more complex sentences. This can be done by enclosing entire subordinate phrases in a rectangle, and hooking this up to the original simple sentence by means of either broken or wavy bonds. Figure 7 shows an example.

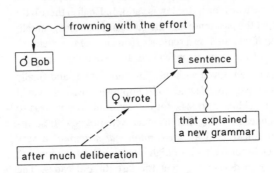

Fig. 7. A more complex case.

It can be seen that the chemist's grammar, thus enlarged, very efficiently warns the user against the errors and obscurities caused by dangling participles. All I have said in Chapters 9 to 12 could be summed up in one simple statement: a subordinate phrase built around a present participle can only be linked to the main sentence through a wavy bond leading from the participle to the male component of the male-female bond.

I hope I have shown that the simple grammar developed in this chapter has its uses, and I also hope to have amused my readers with it. A grammar has an immediate practical use, especially for those who undertake to study a foreign language: it allows the user to determine whether a sentence he has constructed will be acceptable to his intended reader or listener. But this usefulness should not be exaggerated: a grammar does not *generate* sentences, it only *tests* the sentences that have been generated by a far superior mental process. The analogy to chemistry is easy to draw: the synthesis comes first, the purity tests come afterwards.

So the excuse I have for inflicting the present linguistic disquisition on the reader is not its "usefulness", the puny gain obtained by knowing how to use the verb *to react* properly. My real purpose in writing this chapter is to get the reader to have a fresh, uninhibited, look at sentence construction, and to perceive that it is controlled by laws as logical and as interesting as those that govern chemical reactions.

During my work as editor I occasionally find it necessary to build the model of a molecule. It may be that I have to intervene in a dispute on chirality, or I may simply be searching for a better way to picture the molecule in a diagram. I confess I take a sly pleasure in this activity: perhaps it makes me feel like a little boy among his building blocks, or more likely it allows me to dream myself away from my desk-bound job back to the laboratory, the world of the "real" molecules. The model grows in my hands, and for a while all is confusion as the rings twist in and out of boat and chair shapes. But then the last bond snaps into place with a most satisfactory click, and suddenly all is order, symmetry, stability. If your sentences snap into place with a similar click, dear reader, then you are well on the way to writing good prose — and enjoying yourself in the process.

35 *Lights! Camera! Action!*

In the preceding chapter, I introduced a "chemist's grammar" designed to give scientists a better insight into sentence structure. I should now like to present yet another plaything of mine, a grammar with some fascinating features. First of all, it is a universal grammar rather than one that describes the laws of a specific language. Second, unlike a conventional grammar, it makes some statements about efficient communication and even writing style.

My train of thoughts was set in motion when I was planning Chapter 6, on the "German vice". You may recall that this alleged German aberration, the *Hauptwortkombinationenzusammensetzungsbedürfnis,* is actually shared by English, where *noun combination assembly mania* follows the same apparently inefficient word order. In these chains of nouns, each link qualifies the one that follows, so that you have to advance to the end of the chain and then work your way backwards in order to deduce the meaning. Surely such a way of proceeding is an impediment to clear communication, and worthy of the strongest *communication clarity impediment condemnation.* Note the gap that has been created between *strongest* and *condemnation;* the adjective is like a dog that vainly sniffs at three men before finding its master.

Efficiency of communication is hard, but not impossible, to quantify. In Chapter 18, I defined that version of a message as the most efficient that could be understood by the reader in the shortest possible time. Communication with computers can be assessed for efficiency in a similar fashion: if you have to issue an instruction to a computer, then you would prefer the program that takes the smallest number of steps to issue the instruction. It is obvious that the sausage-words that have just set our teeth on edge fail such efficiency tests miserably: after every component noun there follows the program step "store in memory till next noun arrives". Such program steps would be avoided if the order of nouns were reversed.

I was thus greatly tempted to follow several texts on good writing, and to condemn noun chains as "contrary to logic" and as "unnatural". But an incident caused me to think again. I was going for a walk with a lively $2\frac{1}{2}$ - years-old boy. Suddenly, to the delight of my companion, there was heard a most impressive howling of sirens; a firetruck raced up and stopped at a house just ahead of us; and several splendidly helmeted firemen rushed into the house. Young Kym was greatly pleased with the spectacle the Melbourne Fire Authority had turned on for him: he pointed an excited finger at the gleaming helmets and said "The Hat Man come".

Hat man: That set me thinking. Here was this child, not yet three years old, from a good family, not yet corrupted by English teachers – and yet already in the grip of the German vice. He had taken two nouns and linked them together in such a way that the first qualified the second. So much for this device of language being "unnatural". You will recall that my article about the German vice was written in moderate terms. I warned against excessive manufacture of sausage-words, but I also said that in its simplest form, when two nouns are joined together, the device was not to be despised and was in fact an asset to the language.

Then I had another thought. My young friend had said *hat man*. But a French child of the same age would have put the words the other way round: *homme-chapeau,* man-hat. I decided to look up what linguists knew about such matters. It turns out that they use an interesting technical name, only recently coined. Languages such as English or German, in which the qualifying term precedes the qualified, are called "left-branching" because, if you imagine the language written from left to right, you first have to leap to the end of the chain and then branch to the left. The opposite order, as in French, is called "right-branching". (Semitic languages are of course written from right to left, but to the linguist they are right-branching.)

Now it is an extraordinary fact that the human race is about evenly divided between left-branchers and right-branchers. Noun chains such as those in English and the other Germanic languages are also formed in Chinese, Japanese and Turkish. That makes a lot of adherents to an apparently illogical word order!

A particularly amusing example of how ingrained these branching habits are is provided by New Guinea pidgin. This extremely useful language is formed by the union of an English vocabulary and Melanesian grammar, and noun chains in Melanesian are (to the best of my knowledge) right-branching. Thus, if it should rain in Port Moresby, an Australian inhabitant of that city would look for his *raincoat,* a New Guinean for his *kot ren.*

And so the dance of words continues. Of course most languages allow a flexible word order, for emphasis, as in the slighting: *"His results* I wouldn't trust within 10%". But there are languages (the majority) where the subject noun precedes the verb and others where it follows. Even prepositions join the dance: the Hungarian equivalent of *in Washington* works out as *Washington-in.*

But let us talk of adjectives. All the languages that are left-branching for noun chains remain so for adjectives, i.e. they put them in front of the noun, and this large group is now also joined by the Slav languages. In the Latin languages the adjective generally follows the noun but occasionally it is shifted forward, with a slight change in meaning: *une certaine prédiction* is not quite the same as *une prédiction certaine.*

Why? It should be admitted that in some languages there is a practical reason for such word order. Whenever in English a form of *to be* would be the main verb of a sentence, it is simply omitted in Chinese or Russian. *Man good* translates into *the man is good,* whereas changing the word order gives an exact equivalent of *good man,* and thus a distinction in meaning.

But surely this alone does not answer the question why such a large slice of humanity should prefer an apparently inefficient word order. Why do we say *pure examplamine* when logic would demand *examplamine pure?* Why do we say *a big man* when we should logically descend from the group to the sub-group by saying *a man big?*

Here my new grammar may be of some help. It operates as follows: You are expected to transcribe the sentence you wish to examine into the camera script for a silent documentary film, and you will find that the most effective order of words conforms to the order in which the image builds up. If you try this on *a big man,* you will find that the film script actually expresses *big* before *man.* If *man* were to come first, this would mean showing a human image without any point of reference. But this is never done in the movies, except as a joke. In a serious film you would create a point of reference first, such as a normal-sized person looking up in astonishment, or a doorframe through which the giant enters stooping. It seems we have to think *big* before we can think *man,* so that the apparently absurd word order is not so absurd after all.

I am aware that the above is not a precise linguistic analysis. It works in some cases but not in others; if we talk about a *grey mouse* we think of the animal first and the color afterwards. Obviously, the human brain can tolerate a certain amount of inefficiency in communication; it can retain an appreciable number of words in the active area of the memory and assemble them in the correct logical order as the sentence progresses. If your computer has only limited storage capacity it becomes very important to write the most efficient program; but if the storage capacity is large you can take liberties.

Thus I am not pretending that my new grammatical device offers an accurate scientific solution to the question of word order, but it certainly helps us to understand certain apparent departures from logic better. By putting the adjective in front of the noun the speaker or writer creates a sense of expectancy; he causes the noun to make "a more dramatic entrance on the stage". And if this is an advantage in 50% of all cases only it would still explain why half the world's languages put the adjective first.

With the aid of the device one can even explain, although one certainly cannot condone, some of the craziest excesses of the German vice. Take the case of "cathode ray oscilloscope display screen control switchboard operator". To express this in a silent film you would in fact follow the order of the words. You would begin with a medium shot in which the frame would be filled by a cathode ray oscilloscope. The camera clearly sees this as one word, and so do we. We would then track forward until we have the display screen − again one word − in focus. Then we would pan down to show the control switchboard − one word again. And right at the end, the camera would pull back and sidewise to reveal the man who attends the instrument. If we reversed the order of images, we would concentrate attention on the man, whereas it is our intention to show him as the servant of the instrument.

Let me repeat: the above analysis is only meant to explain, not to excuse, the cumbersome word order. It shows us, moreover, that those noun chains are particularly objectionable in which the end overthrows an expectation built up by the beginning. A *word order theory* is acceptable, it gives the impression of smooth camera movement; but a *word order theory refutation* is an atrocious expression.

The "pictorial grammar" thus makes us think about word order, and teaches us much about the nature of language, but I shall not pretend that it is a highly useful tool in determining the best word arrangement. It becomes more and more efficient if we proceed from order within a sentence to order within a paragraph, and then to order among paragraphs. Have we brought the main actors onstage, or have we let the camera dwell on an insignificant detail? Watch the near-silent sequences right at the beginning of big feature films: if the film is made by an experienced director, the background to the story will rapidly emerge, with every foot of film adding an essential detail, and a great expectancy will take hold of the audience. If you can write prose like that, you will stand out among your peers, and your papers will be read by the top men in your field.

But these considerations are too general; let me show off the "pictorial grammar" by putting it to work on a problem that bothers the chemist a good deal. The first two instructions in scientific writing that the young chemist is likely to receive from his supervisor are: "Don't overuse the passive voice" and "Avoid writing in the first person". Now each of these taboos seems reasonable, but taken together they almost cancel out. If we are not allowed to say "we used chromatography" (taboo 2) then we are driven to say in the passive "chromatography was used" (taboo 1). Lately the Chemist's English has become more permissive about the use of the first person, but the young chemist is not quite sure how far he can go. If he really lets himself go and writes "We filtered ... we heated ... then we dissolved ... then we distilled ..." his paper will begin to read, fatally, like the standard schoolboy essay "How I spent my holidays".

To help the writer out of this dilemma, let us first use our camera technique on two everyday sentences. "John crossed the street" is clearly better communication than "the street was crossed by John" because our camera focuses on John and the concept *street* is only introduced by the act of him stepping off one footpath and onto another on the other side. But what if he does not reach the other side? Then it is far better to say, in the passive, "John was hit by a car" because, again, our camera stays with John. If we said, in the active voice, "a car hit John", we would not be telling our story so well because then our camera would follow the car, and we would be leaving poor John behind just when he needed us most.

This sort of reasoning can very profitably be applied to scientific writing. "We assumed examplamine to be tetracyclic because of its mass spectrum" is good writing if there is something noteworthy about the act of assuming; the camera shows excited chemists gathered about a mass spectrum. "Examplamine was assumed to be tetracyclic ..." would then be fuzzy or badly focused camera work: the persons who are making this brave assumption should be in the picture and somehow are not. On the other hand, if the assumption was made by shadowy people in the past, as in "Examplamine was assumed to be tetracyclic, but we found ..." then the original passive is all right: the camera begins in soft focus until *we* appear very sharply in the picture. If, again, there is nothing remarkable about our assuming, then we should not be in the script at all. The sentence should then read "The mass spectrum showed examplamine to be tetracyclic ..." Here the camera shows us a nice mass spectrum and then zooms in on the structural formula and the interesting peaks.

When we come to the experimental section, the technique shows that it is better to say "Examplamine was chromatographed" rather than "We chromatographed examplamine". We the experimenters are only shadowy figures in the background; our silent script is clearly entitled "The perils of examplamine" and in the present episode we see our heroine struggling against the lascivious embrace of the adsorbent. By making examplamine our heroine, we will, if we are clever, be able to keep the passive sentences to a minimum. "Examplamine formed crystals" is far more fun to read than "examplamine was obtained as crystals".

I hope I have convinced the reader that the pictorial grammar has practical uses; perhaps he will be able to think of some further ones himself. Will this device, or any grammatical device, actually help him *construct* an effective sentence? That would be too much to hope for. Grammar is of as much use in sentence construction as a melting point apparatus is in synthesis; it does not help prepare the desired product but is only brought into play afterwards, to check its purity.

This is why this book has not been content with offering the reader precise prescriptions for sentence construction, the "nuts and bolts" of grammar. I have, instead, tried to share with the reader my fascination with language, hoping that he might be seized with equal enthusiasm, and that this enthusiasm might carry him forward to develop his own prose style. And, indeed, our pictorial grammar is a marvelous stimulant for anyone who wants to develop an interest in communication. Our "silent-documentary script" is, in fact, something the theorists of language always wanted: a *metalanguage* so remote from conventional human communication that the properties of language, any language, can be evaluated in it. At the same time, it is a formalized version of the first language all of us learned: the language of images in which a child thinks before he or she develops verbal skills. At the age of 18 months or so, this language enters into competition with the verbal language of the child's ambience, and eventually it is suppressed – but not completely. Highly literate persons such as the hoped-for reader of this book will do nearly their entire thinking in words. But the language in which we all thought in the first year of our life is not lost; it returns to us in moments of tiredness and above all in our dreams.

I write these lines on a Sunday morning. About two hours ago, I woke from a dream so vivid that it has stayed with me, of hurrying down a deserted street and finding all the stores closed. Not a very original dream, surely; it must be common to people of my age who are afraid of approaching retirement and of doors being closed in their faces. Yet – how much structure the dream had! How very powerfully it built up its message of increasing anxiety! And all this was done in the simplest of languages, a language so primitive that I had already at the age of three disdainfully abandoned it for another.

A primitive language, unstructured, with a pitifully small vocabulary, and yet it was perfectly adequate for the message it contained. Can we do less in our papers, with all the glittering resources of the Chemist's English at our disposal? In this book, I have rambled on about all manner of things, picking my subjects not in any logical order but in the order in which I became aware of them in my practice as an editor. Now, as I sit pondering over the last paragraph of the last chapter, it seems to me that a certain structure, a certain underlying order, has emerged almost against my will. I had much to say, at first, about the avoidance of common errors. But the avoidance of errors by itself does not yet generate good style, any more than the mere avoidance of war creates peace. I have returned, in various key chapters, to the theme that your story must always move forward. It is not enough that each sentence fulfill the requirements of the schoolmaster's grammar and the chemist's grammar. It must also create in the reader an expectancy of the next sentence. And each paragraph must create in the reader an expectancy of the next paragraph, each section an expectancy of the next section. Keep your camera steady, your editing taut, move crisply from image to image. At the end, you may get applause from the most discerning of audiences, your colleagues the world over. But your greatest reward will come from the excitement you feel in giving structure to a chapter in your scientific life, and in having attained mastery of the Chemist's English.

Index

[This book is not a highly structured teaching text, so please do not expect every entry in the Index below to point the way to an authoritative treatment of the subject named. My keywords are only meant as memory-joggers, and some have been written with tongue in cheek. I felt it was about time an Index should honor that great Law-giver, Mrs Murphy, who has done so much for science by forcing research workers to overcome unforeseen obstacles.]